Reto Treier

Magnetic Resonance Imaging in Gastroenterology

Reto Treier

Magnetic Resonance Imaging in Gastroenterology

Non-invasive Assessment of Human Gastric Motor Function and Gastric Secretion

Südwestdeutscher Verlag für Hochschulschriften

Impressum/Imprint (nur für Deutschland/ only for Germany)
Bibliografische Information der Deutschen Nationalbibliothek: Die Deutsche Nationalbibliothek verzeichnet diese Publikation in der Deutschen Nationalbibliografie; detaillierte bibliografische Daten sind im Internet über http://dnb.d-nb.de abrufbar.
Alle in diesem Buch genannten Marken und Produktnamen unterliegen warenzeichen-, markenoder patentrechtlichem Schutz bzw. sind Warenzeichen oder eingetragene Warenzeichen der jeweiligen Inhaber. Die Wiedergabe von Marken, Produktnamen, Gebrauchsnamen, Handelsnamen, Warenbezeichnungen u.s.w. in diesem Werk berechtigt auch ohne besondere Kennzeichnung nicht zu der Annahme, dass solche Namen im Sinne der Warenzeichen- und Markenschutzgesetzgebung als frei zu betrachten wären und daher von jedermann benutzt werden dürften.

Verlag: Südwestdeutscher Verlag für Hochschulschriften Aktiengesellschaft & Co. KG
Dudweiler Landstr. 99, 66123 Saarbrücken, Deutschland
Telefon +49 681 37 20 271-1, Telefax +49 681 37 20 271-0, Email: info@svh-verlag.de
Zugl.: Zürich, ETH, Diss., 2008

Herstellung in Deutschland:
Schaltungsdienst Lange o.H.G., Berlin
Books on Demand GmbH, Norderstedt
Reha GmbH, Saarbrücken
Amazon Distribution GmbH, Leipzig
ISBN: 978-3-8381-0402-7

Imprint (only for USA, GB)
Bibliographic information published by the Deutsche Nationalbibliothek: The Deutsche Nationalbibliothek lists this publication in the Deutsche Nationalbibliografie; detailed bibliographic data are available in the Internet at http://dnb.d-nb.de.
Any brand names and product names mentioned in this book are subject to trademark, brand or patent protection and are trademarks or registered trademarks of their respective holders. The use of brand names, product names, common names, trade names, product descriptions etc. even without a particular marking in this works is in no way to be construed to mean that such names may be regarded as unrestricted in respect of trademark and brand protection legislation and could thus be used by anyone.

Publisher:
Südwestdeutscher Verlag für Hochschulschriften Aktiengesellschaft & Co. KG
Dudweiler Landstr. 99, 66123 Saarbrücken, Germany
Phone +49 681 37 20 271-1, Fax +49 681 37 20 271-0, Email: info@svh-verlag.de

Copyright © 2009 by the author and Südwestdeutscher Verlag für Hochschulschriften Aktiengesellschaft & Co. KG and licensors
All rights reserved. Saarbrücken 2009

Printed in the U.S.A.
Printed in the U.K. by (see last page)
ISBN: 978-3-8381-0402-7

Contents

Introduction	3
1 Fast and Optimized T_1 Mapping Technique	13
2 T_1 Mapping for the Quantification of Gastric Secretion	31
3 Influence of Posture on Gastric Function	45
4 Effect of Gastric Secretion on Gastric Physiology	59
Conclusion and Outlook	75
References	83

Introduction

Basic Physiology of Human Gastric Motor Function

Detailed assessment of human gastric motor function is of major importance for a comprehensive understanding of the mechanisms controlling the processing and digestion of food in the gastrointestinal (GI) tract. Food processing and digestion in the GI tract is accomplished by a complex series of various gastric motor activities. Following meal ingestion, the proximal stomach relaxes and accommodates to the meal volume, providing a reservoir for meal storage and enabling an increase in stomach volume while maintaining a constant intragastric pressure. Distension of the stomach and chemical stimulation of gastric cells induce the production of gastric juice by parietal cells (gastric phase of secretion). The principal constituents of gastric juice, hydrochloric acid and pepsin, denature the proteins present in the ingested meal. Gastric contraction waves, which are generated by electrical impulses of the gastric pacemaker at a constant frequency of about three per minute, mix the chyme with secreted gastric juice and grind mechanically large meal particles in the distal stomach (while the pylorus is closed). For short time periods the pylorus opens and small portions of the chyme are emptied into the small intestine. The rate of pyloric opening and closure as well as other physiological changes of the stomach as for example gastric relaxation and accommodation are regulated via complex feedback mechanisms mediated by chemo- and mechanoreceptors located in the small intestine.

Pathophysiology of Human Gastric Motor Function

Disordered gastric motor function is a highly prevalent disease in the Western world. Gastric motility disorders impair seriously the quality of life[1-4] and impose a significant economic burden to both the individual patient as well as to the society as a whole.[5-7] The most common observed gastric motility disorders are gastroparesis, dyspepsia, and gastroesophageal reflux disease (GERD). Gastroparesis is characterized by delayed gastric emptying in the absence of any mechanical outlet obstruction and results in various symptoms such as early satiety, nausea, vomiting, bloating, and upper abdominal discomfort.[8] The main gastric motility abnormalities occurring in dyspeptic patients are delayed gastric emptying, impaired meal distribution in the stomach, impaired accommodation, abnormal fundic contractions, and disturbed antroduodenal motility.[9-15] Patients suffering from dyspepsia feel discomfort or pain in the upper abdomen, often associated with negative symptoms of fullness, bloating, and early satiety.[16,17] GERD is also considered to be a manifestation of a motility disorder, although the symptoms are related predominantly to the effects of hydrochloric acid and pepsin.[18] Most reflux episodes are induced by inappropriate transient lower esophageal sphincter relaxations (TLESR) that are triggered by gastric fundic distension. Typical reflux symptoms are heartburn and regurgitation.[18] The results of different studies indicate that genetic factors modified by environmental factors may be the cause of the described gastric motility disorders;[9,19,20] however, the underlying pathogenesis is still not fully understood.

Standard Clinical Measurement Techniques

A variety of specific measurement techniques exists to analyze gastric motor function in humans. While these techniques provide accurate physiological assessment of gastric motor function, results of several studies have shown their potential value for clinical diagnosis and therapy monitoring of gastric motility disorders. For example, in patients with gastroparesis and functional dyspepsia, gastric emptying of solid meals was significantly delayed and post-prandial gastric volumes were significantly smaller compared to healthy volunteers.[21,22] In GERD patients, the distribution of proximal unbuffered acidic volume ("acid pocket") and its extension onto the distal esophagus was significantly different from that measured in healthy volunteers.[23,24]

Scintigraphy

The gold standard for measuring gastric emptying is scintigraphy.[25] This technique is based on the detection of the emitted gamma radiation of a radiolabeled meal using a gamma camera. Simultaneous assessment of solid and liquid meal emptying is feasible by labeling the solid and liquid meal component with two different radionuclides (mostly ^{111}In and ^{99m}Tc). One major problem concerning scintigraphy is the exposure of the volunteer/patient to ionizing radiation. This limits the application in case of long or repetitive examinations. Furthermore, scintigraphy represents a two-dimensional projection technique providing no information on the origin of the detected signal in the plane perpendicular to the image slice. Due to the complex three–dimensional geometry of the GI tract this is critical, especially if both the stomach and intestines contain portions of the labeled meal.

Breath Test

The stable isotope breath test is another widely used technique in the clinic for the measurement of gastric emptying. Thereby, specific test meals are labeled with the nonradioactive isotope ^{13}C, which is absorbed in the proximal small intestine and metabolized in the liver to $^{13}CO_2$. Breath samples are collected and the expired $^{13}CO_2$ is analyzed by mass spectroscopy. This test assumes that the gastric emptying rate is reflected by the characteristics of the measured breath test curves.[26] However, in patients with liver and pancreatic diseases or malfunctioning intestinal absorption this test clearly fails.

Intraluminal Manometry

Intraluminal manometry is a frequently used technique for the assessment of gastric peristaltic activity. Pressure recordings from the antrum, pylorus, and duodenum can be obtained concurrently by water–perfused catheters. Nevertheless, there are couple of disadvantages. The technique is invasive and time–consuming and it is not possible to quantify geometry–related changes of the travelling contraction waves.

Ultrasound

Ultrasound, as an established medical imaging technique commonly used in clinical routine, allows the visualization and evaluation of gastric peristaltic activity, as well as gastric emptying and gastric accommodation.[27] However, this technique is highly operator–dependent and interpretation of acquired images requires a lot of expertise. Moreover, as with any sonographic technique, the presence of air in the GI tract presents a major limitation for accurate imaging of gastric motor function.

Esophageal pH–Metry

Esophageal pH monitoring is considered the gold standard for the diagnosis of GERD.[28] When combined with intraluminal impedance measurements, also weakly acidic and alkaline reflux events can be detected. For a more detailed understanding of GERD the simultaneous visualization of the anatomical structure of the TLESR as well as the distribution of the gastric acid within the proximal stomach is necessary. This, however, is not feasible using esophageal pH–metry combined with intraluminal impedance measurements.

Magnetic Resonance Imaging

Magnetic resonance imaging (MRI) has been established as a versatile, noninvasive, and ionizing–free medical imaging technique with excellent soft tissue contrast that allows the assessment of human anatomical structures and pathologies as well as cardiac and brain function. MRI for the analysis of human gastric motor function represents a new field of application that is currently performed in only a few research centers world–wide. In the first study fifteen years ago, gastric emptying was the first physiological parameter of gastric motor function that was determined using MRI.[29] More studies followed, and MRI has been shown to be applied successfully for the assessment of gastric accommodation and gastric peristaltic activity.[30,31]

Over the last few years, more and more high–performance MRI systems with increasing magnetic field strength (≥ 1.5 T) became widely available. These high–field MRI systems provide higher signal amplitudes and thus better signal–to–noise ratio (SNR) compared to low–field MRI systems. Excellent image quality is very important in GI–MRI in order to detect small organ structures as for example stomach wall and pylorus. The simultaneous improvement in gradient strength together with advanced imaging techniques (sensitivity encoding (SENSE);[32] k–t BLAST and k–t SENSE[33]) speeds up considerably data acquisition time. Because of the irregular gastric and intestinal peristaltic activity fast data acquisition is the crucial factor in GI–MRI to generate artifact–free images.

Most of these new high–field MRI systems are built in compact architecture allowing imaging only in the lying body position. This is not problematic for the investigation of cardiac and brain function, but has to be considered in GI–MRI. In a variety of scintigraphic studies it has been shown that posture has a significant influence on human gastric motor function.[34-38] In order to establish MRI in GI research and in clinical routine for the assessment of gastric motility disorders in the future, detailed investigation of the global effect of posture on gastric motor function is essential.

Paramagnetic Contrast Agents

Paramagnetic contrast agents have been widely used in gastric MRI to qualitatively assess the distribution of oral drug delivery systems and of solid/liquid meals in the gastric lumen.[39-41] Basically, paramagnetic contrast agents affect the MRI relaxation properties of the labeled meal or perfused tissue (if administered intravenously). Two different relaxation effects are observed, a transverse or T_2 relaxation and a longitudinal or T_1 relaxation. After excitation using a radiofrequency (RF) pulse, spins of the water protons tumble around the magnetic field. Since the actual magnetic field is a superposition of the static magnetic field B_0 and local, time–varying electromagnetic fields induced by the magnetic dipole moments of neighbouring protons, each spin tumble with a slightly different precession frequency. This causes a dephasing of the spin system and results in a decrease in the amplitude of the macroscopic magnetization (transverse relaxation). The rate of dephasing due to these spin–spin interactions is described by the T_2 relaxation time. However, in MRI additional linear gradient fields are applied for spatial encoding of the signals. These gradient fields (as well as B_0 inhomogeneities) cause an accelerated dephasing of the spin systems which is described by the time constant T_2^* ($T_2^* < T_2$). Simultaneously to dephasing, excited protons interact with local, time–varying electromagnetic fields induced by a combination of translations, rotations, and vibrations in the molecular environment (the "lattice"). If the frequency of these fields is close to the Larmor frequency $\omega_L = \gamma \cdot B_0$, where γ is the gyromagnetic ratio of the proton, energy is transferred to the lattice in the form of thermal energy. This causes the macroscopic magnetization to relax back to its thermal equilibrium state at a rate described by the T_1 relaxation time (longitudinal relaxation).

Paramagnetic contrast agents shorten both T_1 and T_2 (T_2^*); however, typical applications in MRI specifically rely on shortening of T_1. The longitudinal relaxation rate depends on the magnitude of the contrast agent's electronic magnetic moment. Since the magnetic moment of a particle is proportional to its gyromagnetic ratio and the square of its spin, elements with unpaired electrons ($\gamma_{electron} = 657 \cdot \gamma_{proton}$) and high total spin are effective T_1 relaxation agents. In addition, the longitudinal magnetization rate depends also on the electron spin relaxation time, a measure of how long the magnetic dipole of the electron remains in its own spin state. Elements with the highest numbers of unpaired electrons and the longest electron spin relaxation times will have the strongest influence on longitudinal relaxation rate of protons. With seven unpaired electrons in the 4f valence shell and a long electron spin relaxation time of $10^{-8} - 10^{-9}$ s, the Gadolinium (Gd^{3+}) ion is the optimum choice.[42]

Unfortunately, Gadolinium is toxic in the human body preventing the administration in its elemental form. Therefore, Gd^{3+}–ions are bound to chelate complexes as for example diethylene triamine pentaacetic acid (DTPA; Figure 1a) or tetraazacyclododecane tetraacetic acid (DOTA; Figure 1b). These chelates remain stable in the human body until they are excreted, even if they are exposed to a highly acid solution as gastric juice.[29] Although some of the unpaired electrons of the Gd^{3+}–ion are now paired with the chelate molecule and the minimal distance between water protons and the Gd^{3+}–ion is increased, Gd–DTPA and Gd–DOTA decreases significantly the longitudinal relaxation time T_1 of the labeled meal.

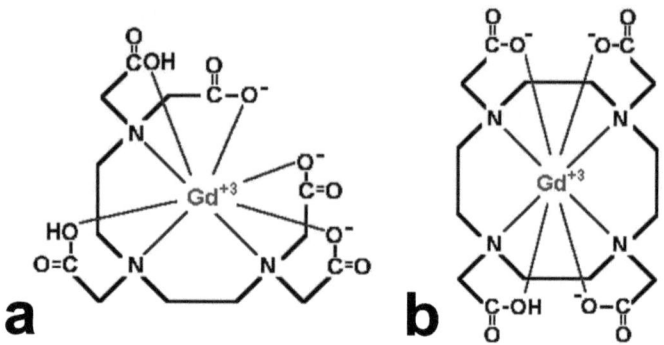

Figure 1: Schematic representation of the molecular structures of Gd–DTPA (a) and Gd–DOTA (b), respectively.

Gd–based paramagnetic contrast agents applied in diagnostic radiology are available as diluted solutions. For the assessment of the distribution process of solid/liquid meals in the gastric lumen, the liquid meal component is homogeneously mixed with a few milliliters of the liquid contrast agent before meal ingestion. On the MR images signal intensity of the labeled liquid meal component is enhanced and can qualitatively be separated from the signal intensity of the solid meal component, intragastric air, and stomach wall using an operator–defined threshold. Since paramagnetic contrast agents decrease the longitudinal relaxation time T_1, these effects are even more pronounced by applying T_1–weighted MRI sequences. For accurate image analysis based on the comparison of signal intensities, a homogeneous excitation of the spins over the entire abdomen is required. However, especially for high–field MRI systems, the transmitted RF field is extremely inhomogeneous. Furthermore, this analysis method is highly operator–dependent and therefore only qualitative.

Measuring the T_1 relaxation time of the labeled liquid meal presents a new approach allowing to quantitatively assess distribution processes in the gastric lumen. As described in literature[43] there is an interrelationship between the concentration C of the contrast agent and T_1 of the labeled liquid meal

$$\frac{1}{T_1(C)} = \frac{1}{T_1(C=0)} + r_1 \cdot C,$$

where r_1 represents the relaxivity of the contrast agent. By dynamically measuring T_1 and calculating the corresponding concentration C, distribution, dilution, and mixing processes in the GI tract can be quantified.

T_1 Mapping Techniques

Inversion Recovery and Saturation Recovery

Inversion recovery (IR) and saturation recovery (SR) represent the two standard MRI sequences for T_1 determination. The basic pulse sequence is shown schematically in Figure 2. After a spin preparation pulse (inversion (180°) pulse for IR and saturation (90°) pulse for SR) followed by a delay time T_D, an excitation pulse (90° in this example) is applied and the free induction decay (FID) or echo signal is sampled. According to the Bloch equations the signal amplitude S is proportional to the delay time T_D and relaxation time T_1:

$$S \propto \left(1 - c \cdot \exp\left(-\frac{T_D}{T_1}\right)\right),$$

with $c = 2$ for IR and $c = 1$ for SR.[43]

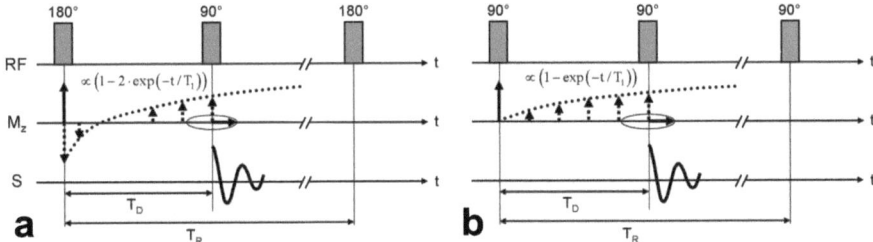

Figure 2: Radiofrequency (RF) pulse sequence, relaxation curve of the longitudinal magnetization (M_z), and free induction decay signal (S) of inversion recovery (a) and saturation recovery (b). The delay time is defined by T_D and the repetition time by T_R.

This basic sequence is repeated consecutively using different delay times T_D and with a repetition time $T_R \geqslant 5 \cdot T_1$ in order to allow nearly full recovery (> 99 %) of the longitudinal magnetization before the next preparation pulse is applied. The T_1 relaxation time is derived using a two–parameter (a, b) nonlinear least squares fit of the form $a \cdot \left(1 - c \cdot \exp\left(-\frac{t}{b}\right)\right)$ to the measured signal amplitudes (with $c = 2$ for IR and $c = 1$ for SR). In case of inaccurate preparation pulses resulting in flip angles different from 180° and 90°, respectively, a three–parameter nonlinear least squares fit must be applied with the fitting parameters (a, b, c). IR and SR techniques are easy to apply on every clinical MRI system without sequence programming but have the disadvantage, because of their long acquisition times, of being very sensitive to respiratory and organ motion. Furthermore, these techniques are not feasible to assess dynamics of contrast agent distribution as for example in perfusion measurements.

Look–Locker Technique

Modified IR and SR sequences based on the theory of Look and Locker[44] have much shorter acquisition times resulting in a clinically relevant time resolution in the order of second per T_1 map. Basically, after a spin preparation pulse the recovery curve of the longitudinal magnetization is sampled by the acquisition of a series of images within one T_R. The Look–Locker technique can be combined with any fast readout technique (echo planar imaging (EPI);[45] T_1–weighted fast field echo (T_1–FFE);[46] balanced fast field echo (bFFE)[47]). Since the applied excitation pulses during readout alter the time evolution of the longitudinal magnetization, modified equations must be used to describe the characteristics of the recovery curve.[45–47] T_1 is calculated pixel–by–pixel using a multiple–parameter nonlinear least squares fit to the measured signal amplitudes.

Variable Flip Angle Technique

In addition to the Look–Locker sequence there is another common fast T_1 mapping technique which is based on the variable flip angle (VFA) approach. The basic principle of the VFA approach first introduced by Fram et al.[48] in 1987 is completely different from that of the Look–Locker technique. Instead of sampling the recovery curve, several images are acquired by applying repeated (at least two) T_1–FFE sequences with different flip angles. The signal intensity of a T_1–FFE sequence is a complex function of the equilibrium longitudinal magnetization, relaxation times T_1 and T_2, and sequence parameters (echo time, repetition time, and flip angle). Measuring the signal intensities using different flip angles but identical sequence parameters allows the calculation of T_1 values pixel–by–pixel.[48]

Objectives and Outline

The aim of the presented thesis was to develop, evaluate and apply dedicated MRI techniques to analyze human gastric motor function and its dependency on body position as well as to quantify gastric secretion.

The methodological, first part (Chapters 1 & 2) introduces the development and evaluation of a fast T_1 mapping technique based on the variable flip angle approach for the quantitative assessment of gastric secretion. Optimization methods, as described in Chapter 1, were developed to achieve maximal accuracy in T_1 determination within a limited acquisition time (2.3 s per T_1 map). In Chapter 2, this fast and optimized T_1 mapping technique for the noninvasive quantification of gastric secretion was evaluated in healthy volunteers. Basically, gastric secretion was assessed by measuring changes in the relaxation time T_1 of an ingested liquid test meal homogeneously labeled with the paramagnetic contrast agent Gd–DOTA of known concentration. Using the interrelationship between T_1 and contrast agent concentration, which was derived in an *in vitro* experiment, changes in concentration and thus volume and distribution of gastric secretory products were determined.

In the second part (Chapters 3 & 4), two studies are presented aimed on investigating the influence of posture on gastric motor function and the physiological response to stimulated gastric secretion. In Chapter 3, an imaging protocol is described allowing the simultaneous assessment of eight relevant parameters of gastric motor function (stomach, meal, and intragastric air volume; gastric relaxation and emptying; intragastric meal distribution; peristaltic frequency and velocity). Protocols were applied on two MRI systems of different architecture (whole–body vs. open–configuration). Thereby, the overall effect of the lying body position on gastric motor function was analyzed. In a physiological study, as presented in Chapter 4, the fast and optimized T_1 mapping technique was applied to investigate the effect of stimulated gastric secretion on gastric volume responses, emptying and intragastric dilution. Twelve healthy volunteers were studied on two different days with either pentagastrin (as stimulus for gastric secretion) or placebo (NaCl solution) administered intravenously in double–blind randomized order.

Chapter 1

Optimized and Combined T_1 and B_1 Mapping Technique for Fast and Accurate T_1 Quantification in Contrast–Enhanced Abdominal MRI

Introduction

Fast and accurate T_1 mapping techniques allow the quantitative assessment of the distribution and dynamics of intravenously or orally applied paramagnetic contrast agents (CAs). Clinical MRI applications, such as perfusion and permeability measurements, mainly use T_1–weighted dynamic contrast–enhanced imaging (DCE–MRI) sequences to detect and characterize lesions and tumors, monitor therapy,[49–51] and assess organ function in general.[52–54] Similar approaches are applied in gastrointestinal research to evaluate distribution and emptying processes in the human stomach.[39,55] With these techniques, parameters of interest can be assessed in a semiquantitative way by detecting differences in signal intensity that are induced by changes in CA concentration.[39,55,56] However, such semiquantitative analyses present a major drawback because signal intensity not only reflects CA concentration but also depends on intrinsic tissue properties (T_1, T_2, and spin density ρ), imaging parameters (e.g., the flip angle (FA), repetition time T_R, and echo time T_E), and the hardware of the MRI scanner (e.g., receiver gain), thus preventing standardization of the DCE–MRI technique.[50,51] A direct quantitative assessment of change in CA concentration can be achieved by dynamically measuring the T_1 relaxation time and using the interrelationship

$$\frac{1}{T_1} = \frac{1}{T_{10}} + r_1 \cdot C, \qquad (1.1)$$

where r_1 is the relaxivity and C is the concentration of the CA.[43] However, because of their long acquisition times, standard MRI sequences for T_1 quantification involving partial saturation and inversion recovery (IR) are not feasible for resolving such dynamic processes at a sufficiently high temporal resolution. Furthermore, these sequences show high sensitivity to respiratory and organ motion.[57]

The aim of this study was to optimize a fast T_1 mapping technique for accurate T_1 quantification in abdominal CE–MRI using the variable FA (VFA) approach.[58,59] Optimization methods were developed to maximize the signal–to–noise ratio (SNR) and ensure effective spoiling and steady state for a defined T_1 range and a limited T_R and acquisition time. Similar methods were applied to optimize a fast B_1 mapping technique required to correct for spatial variations of the nominal FA. By combining the optimized T_1 and B_1 mapping techniques, high–precision abdominal T_1 maps of healthy volunteers were generated in a limited acquisition time (2.3 s per T_1 map).

Theory

The VFA technique is based on the consecutive application of T_1–weighted spoiled gradient–echo (T_1 fast–field echo (T_1–FFE)) sequences using different FAs.[48] The theoretical signal intensity S of a T_1–FFE sequence is a function of the equilibrium (longitudinal) magnetization M_0, relaxation times T_1 and T_2, echo time T_E, repetition time T_R and FA α:

$$S = M_0 \cdot \frac{\sin\alpha \cdot (1 - E_1) \cdot E_2}{1 - E_1 \cdot \cos\alpha}, \tag{1.2}$$

where $E_1 = \exp\left(-\frac{T_R}{T_1}\right)$ and $E_2 = \exp\left(-\frac{T_E}{T_2}\right)$.[43]

Equation (1.2) can be expressed in a linear form $y = a \cdot x + b$:

$$\frac{S}{\sin\alpha} = E_1 \cdot \frac{S}{\tan\alpha} + M_0 \cdot (1 - E_1) \cdot E_2. \tag{1.3}$$

Keeping T_R and T_E constant and measuring the signal induced by different FAs, a straight line characterized by slope $a = E_1$ and intercept $b = M_0 \cdot (1 - E_1) \cdot E_2$ can be found by linear regression. T_1 is then easily calculated as:

$$T_1 = -\frac{T_R}{\ln a}. \tag{1.4}$$

Equation (1.2) and thus a correct calculation of T_1 are only valid if the transverse magnetization is completely spoiled, longitudinal magnetization has reached a dynamic equilibrium (steady state), and the amplitude of the radiofrequency field (B_1) is homogeneous over the complete field of view (FOV). Especially for a large FOV and at high magnetic field strengths, B_1 field inhomogeneities represent a dominant source of error in the VFA approach.[60]

Materials and Methods

As described in the previous paragraph, several sequence requirements, including effectual spoiling, steady state, and B_1 field homogeneity, must be fulfilled for precise T_1 mapping using the VFA approach. Furthermore, a sufficiently high SNR is also needed for accurate T_1 mapping. In the following paragraphs, optimization methods for maximizing SNR and ensuring effective spoiling and steady state are described for a defined T_1 range of 100 – 800 ms and a limited T_R (9 ms) and acquisition time. In addition, an optimized fast FA imaging (B_1 mapping) technique[61] to correct spatial variations of the nominal FA is presented.

Image Acquisition

Optimal FAs

T_1 determination using the VFA approach requires a minimum number of two consecutive measurements with two different FAs α_1 and α_2 (and corresponding signals S_{α_1} and S_{α_2}).[48] The two optimum FAs are derived by maximizing the product of the normalized dynamic range (DR) times the fractional signal (FS).[62] DR is given by

$$DR = \frac{S_{\alpha_1}}{M_0 \cdot \sin \alpha_1} - \frac{S_{\alpha_2}}{M_0 \cdot \sin \alpha_2}, \qquad (1.5)$$

describing the normalized separation of the two data points along the computed regression line. FS is defined as

$$FS = \frac{S_{\alpha_1} + S_{\alpha_2}}{2 \cdot S_{\alpha_E}}, \qquad (1.6)$$

where S_{α_E} stands for the maximum possible signal after excitation using the Ernst angle $\alpha_E = \arccos(E_1)$. The product $DR \cdot FS$ is a function of four parameters: T_R, T_1, α_1, and α_2 (assuming $E_2 \approx 1$ for $T_E \ll T_2$). For $T_R = 9$ ms and for every FA combination (S_{α_1}, S_{α_2}), $DR \cdot FS$ is integrated over the defined T_1 range of 100 – 800 ms. The integrated product reaches its maximum for the optimum FAs $\alpha_{1_{opt}} = 5°$ and $\alpha_{2_{opt}} = 31°$.

SNR

3D volumetric excitation and signal averaging provide a means of improving the SNR, but have the disadvantage of increasing the acquisition time.[43] Therefore, in order to minimize the acquisition time per image slice, a multiple 2D excitation technique with no signal averaging was applied in this study. SNR was optimized by applying radiofrequency (RF) excitation pulses with a high specific bandwidth (product of bandwidth times pulse duration). These high–quality RF excitation pulses provide improved rectangular slice excitation profiles and hence improved SNR compared to standard RF excitation pulses. A sinc–Gaussian RF excitation pulse with a specific bandwidth of 137 rad is used throughout the measurements.

RF and Gradient Spoiling

Suppression of higher–order echoes is achieved by linearly incrementing the phase ϕ of consecutive RF excitation pulses (RF spoiling) and applying strong crusher gradients in the readout and slice–selection directions (gradient spoiling). For optimal RF spoiling over a wide range of T_1, T_2, and T_R, a phase increment of $\Delta\phi = 150°$ is applied.[63] Gradient spoiling is optimized by maximizing the product of gradient strength times gradient duration of both crusher gradients. The crusher gradients are applied at the end of signal acquisition within the given T_R, and the gradient amplitude is set to the highest possible value allowed by the hardware.

Steady State

For T_1–FFE sequences, the steady state magnetization (i.e., equal longitudinal magnetization at the time of each excitation) is established after a certain number of dummy excitations. The number of dummy excitations depends on T_R, T_1, and FA α, and can be found by iterative calculation using

$$M_z(n) = (M_z(n-1) \cdot \cos\alpha - M_0) \cdot exp\left(-\frac{T_R}{T_1}\right) + M_0, \quad (1.7)$$

where $M_z(n)$ represents the longitudinal magnetization before the n–th excitation.[43]

Figure 1.1a and Figure 1.1b show the time evolution of the longitudinal magnetization for two consecutive measurements using $\alpha_{1_{opt}}$ and $\alpha_{2_{opt}}$, and for different T_1 values in the range of 100 – 800 ms. For the first measurement the initial longitudinal magnetization $M_z(0)$ corresponds to the equilibrium longitudinal magnetization M_0 (Figure 1.1a). The initial longitudinal magnetization of the second measurement corresponds to the steady–state magnetization of the first measurement (Figure 1.1b). Therefore, in order to minimize the total duration to reach steady state, the lower optimal FA is applied first. Further reduction of scan duration is obtained by applying an optimized initial FA for each measurement. There is only one characteristic initial FA for each T_1 value (and for a fixed T_R of 9 ms) that establishes the steady–state condition after one single RF excitation pulse. These characteristic initial FAs are determined for T_1 values of 100 – 800 ms, and range from 16° (for T_1 = 100 ms) to 41° (for T_1 = 800 ms) for $\alpha_{1_{opt}}$ and from 70° (for T_1 = 100 ms) to 87° (for T_1 = 800 ms) for $\alpha_{2_{opt}}$. By simulating the corresponding time evolution of the longitudinal magnetization, the maximum number of excitations to reach the steady state for each initial FA within the given T_1 range can be determined. The minimum over all these maxima defines the optimized initial FAs, which are 33° for $\alpha_{1_{opt}}$ (Figure 1.1c) and 81° for $\alpha_{2_{opt}}$ (Figure 1.1d).

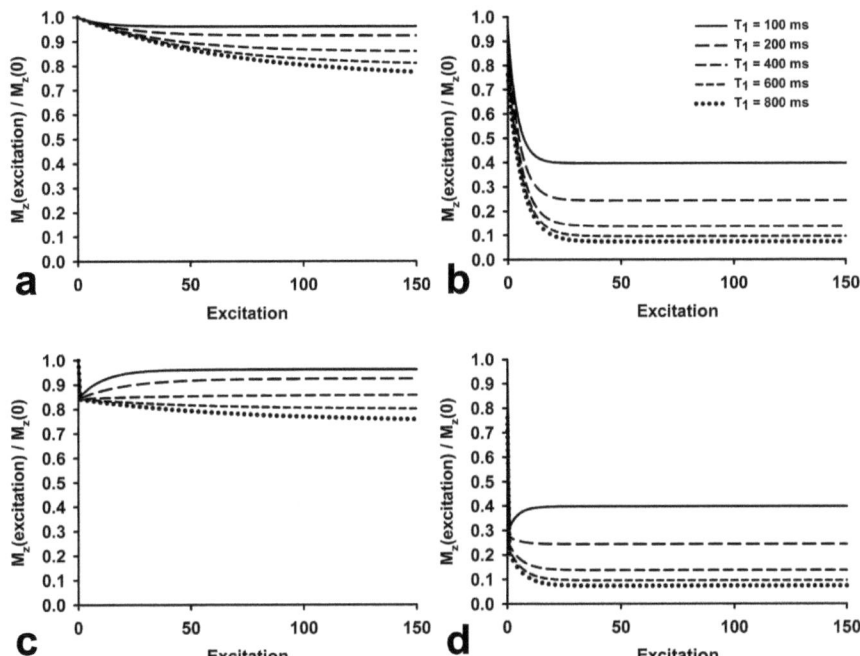

Figure 1.1: *Relative longitudinal magnetization as a function of excitation for two consecutive measurements using $\alpha_{1_{opt}}$ (a) and $\alpha_{2_{opt}}$ (b). The altered relative longitudinal magnetization using the optimized initial FAs ((c) 33° and (d) 81°) prior to $\alpha_{1_{opt}}$ and $\alpha_{2_{opt}}$ is shown. The duration to establish the steady state over the complete T_1 range is minimized when the optimized initial FAs are applied. All curves are plotted for $T_R = 9$ ms and $T_1 = [100\ ms,\ 200\ ms,\ 400\ ms,\ 600\ ms,\ 800\ ms]$.*

For *in vitro* and *in vivo* experiments the steady–state condition is assumed for

$$\left| \frac{M_z(n) - M_z(n-1)}{M_0} \right| < 0.001, \tag{1.8}$$

i.e., when the relative difference of the longitudinal magnetization before two consecutive excitations is less than 0.1 %. This results in 29 dummy excitations for $\alpha_{1_{opt}}$ and optimized initial FA of 33° before actual data acquisition and in 18 dummy excitations for $\alpha_{2_{opt}}$ and optimized initial FA of 81°.

Spatial variations of the nominal FA are corrected by performing an additional measurement applying an optimized B_1 mapping technique. This technique is based on a T_1–FFE sequence with a fixed FA α and two alternating repetition times ($T_{R_1} < T_{R_2}$).[61] The ratio r of the observed gradient–echo signals $S_{T_{R_1}}$ and $S_{T_{R_2}}$ is a function of α, T_{R_1}, T_{R_2}, and T_1:

$$r = \frac{S_{T_{R_1}}}{S_{T_{R_2}}} = \frac{1 - E_{12} + (1 - E_{11}) \cdot E_{12} \cdot \cos\alpha}{1 - E_{11} + (1 - E_{12}) \cdot E_{11} \cdot \cos\alpha}, \tag{1.9}$$

where $E_{11} = \exp\left(-\frac{T_{R_1}}{T_1}\right)$, and $E_{12} = \exp\left(-\frac{T_{R_2}}{T_1}\right)$.

Based on this equation, the FA α is calculated as

$$\alpha = \arccos\left(\frac{1 - r \cdot \left(\frac{1-E_{11}}{1-E_{12}}\right)}{r \cdot E_{11} - \left(\frac{1-E_{11}}{1-E_{12}}\right) \cdot E_{12}}\right). \tag{1.10}$$

If the first–order approximation can be applied to exponential terms, $\exp\left(-\frac{T_R}{T_1}\right) \approx 1 - \frac{T_R}{T_1}$, Equation (1.10) simplifies to

$$\alpha = \arccos\left(\frac{1 - r \cdot n}{r \cdot \left(1 - \frac{T_{R_1}}{T_1}\right) - n \cdot \left(1 - \frac{T_{R_2}}{T_1}\right)}\right), \tag{1.11}$$

where $n = \frac{T_{R_1}}{T_{R_2}}$.

Assuming that $1 - \frac{T_{R_1}}{T_1} \approx 1$ and $1 - \frac{T_{R_2}}{T_1} \approx 1$ for $T_{R_1}, T_{R_2} \ll T_1$ the calculation of the FA becomes independent of T_1 and is only a function of the ratios of the measured signals r and the repetition times n:

$$\alpha = \arccos\left(\frac{1 - r \cdot n}{r - n}\right). \tag{1.12}$$

The sensitivity $\frac{dr}{d\alpha}$ of the FA determination using Equation (1.12) increases as the FA α and the factor $\frac{1}{n}$ increases (Figure 1.2a). Obviously, an increase in the duration of T_{R_1} and T_{R_2} also improves the sensitivity, but has the disadvantage of prolonging the scan duration. In this study the optimized sequence parameters $T_{R_1} = 20$ ms, $T_{R_2} = 100$ ms and $\alpha = 70°$ are chosen for B_1 mapping. This results in a maximal error in FA determination of < 4.2 % when Equation (1.12) is applied over a T_1 range of 100 – 800 ms (Figure 1.2b).

RF and gradient spoiling is optimized in the same way as described for the T_1 mapping technique. The phase of consecutive excitation pulses is linearly incremented by 150°, and the highest possible amplitude of both crusher gradients

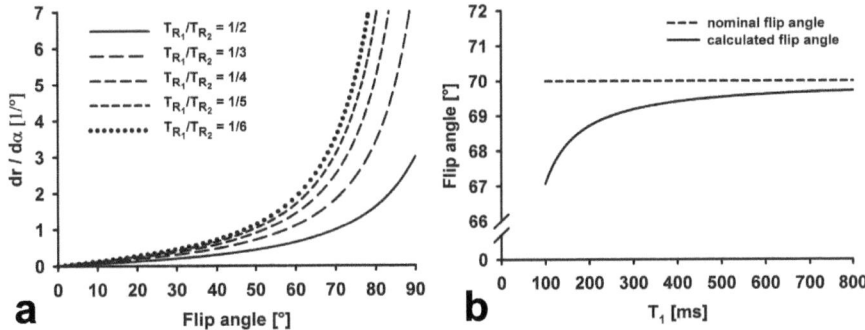

Figure 1.2: **(a)** *Sensitivity $dr/d\alpha$ as a function of the FA plotted for different ratios T_{R_1}/T_{R_2}. For high FAs and a low ratio T_{R_1}/T_{R_2} the sensitivity of FA determination is markedly increased.* **(b)** *Calculated FA over T_1 range of 100 – 800 ms using optimized sequence parameters $T_{R_1} = 20$ ms, $T_{R_2} = 100$ ms and $\alpha = 70°$. The nominal FA is well approximated by the calculated FA, especially for high T_1 values. The maximum deviation from the nominal FA is 4.2 % for $T_1 = 100$ ms.*

is applied in the readout and slice–selection directions. Six dummy excitations are used to establish the steady–state condition as defined in Equation (1.8). The identical RF excitation pulse, image geometry, and RF power are used for both the T_1 mapping and subsequent B_1 mapping measurements, resulting in equal FA distribution over the complete FOV.

Experiments

In vitro and *in vivo* measurements were performed on 1.5 T and 3 T whole–body MRI systems (1.5 T Achieva and 3 T Achieva, Philips Medical Systems, Best, The Netherlands) using an abdominal, four–channel phased–array receive coil. The B_1 mapping measurement was always performed immediately after the T_1 mapping measurement. A linear profile order was used for readout in k–space. The imaging parameters for both MRI systems (specific to 3 T in brackets) were as follows: FOV = 360 mm, slice thickness = 15 mm; T_1 *mapping*: FAs = 5° and 31°, initial FAs = 33° and 75° (33° and 45°), number of dummy excitations = 29 and 21, scan matrix = 128×128, T_R/T_E = 9/3.6 ms (23/4.6 ms), scan time = 2.3 s per T_1 map (5.8 s per T_1 map); B_1 *mapping*: FA = 70°, number of dummy excitations = 6, scan matrix = 64×64, T_{R_1} = 20 ms (23 ms), T_{R_2} = 100 ms, T_E = 3.6 ms (4.6 ms), scan time = 6.6 s per b_1 map (6.8 s per b_1 map). The optimized initial FA of 81° for $\alpha_{2_{opt}}$ was not applicable at 1.5 T and 3 T due to hardware constraints (duty cycle).

In Vitro Experiments

Six concentrations (1200 µM, 1000 µM, 800 µM, 600 µM, 400 µM, and 200 µM) of the paramagnetic CA Gd–DOTA (DOTAREM®, Laboratoire Guerbet, France) homogeneously mixed with a viscous solution of 10 % glucose and 1 % locust bean gum powder (LBG) were prepared in small bottles (100 ml) at room temperature. The bottles were simultaneously placed in the isocenter of the 1.5 T and 3 T MRI system, respectively, and one image slice aligned perpendicular to the long axes of the bottles was acquired for T_1 determination. T_1 values (mean (standard deviation (SD))) were calculated within a region of interest (ROI) of 6×6 pixels for each bottle. Furthermore, a reference IR spectroscopy measurement ($T_R = 3$ s, six inversion delay times equally spaced by 10 ms and appropriately chosen to detect the zero–crossing of the signal amplitude in the spectrum) was performed for each concentration at room temperature and additionally for the 1200 µM concentration at a temperature of 37°C. T_1 was determined by detecting the zero–crossing of the signal amplitude at time point τ and using the interrelationship $T_1 = \tau/ln2$.

In Vivo Experiments

Two healthy volunteers were investigated for the *in vivo* experiments. Written informed consent was obtained from each volunteer and the protocol was approved by the local ethics committee. The sequence was validated on the 1.5 T MRI system in one healthy volunteer after a 6–hr fast. A small balloon attached to the end of a nasogastric tube was intubated nasally and positioned within the proximal part of the stomach. The balloon was filled with 300 ml of a 10 % glucose solution homogeneously mixed with 1 % LBG and 1200 µM Gd–DOTA at 37°C (reference T_1 value: 202 ms (4 ms)). For subsequent imaging the volunteer was positioned in supine body position in the MRI system. Four oblique transverse image slices covering the proximal part of the stomach were acquired within one breath–hold for both the T_1 mapping (total scan time = 9.2 s) and consecutive B_1 mapping (total scan time = 26.4 s) measurements. T_1 values (mean (SD)) of the ingested solution were determined for each T_1 map. Another T_1 mapping measurement was performed on the 3 T MRI system in a different healthy volunteer after 6–hr fast (without intubation). The ingested reference solution, the scan protocol, and the data analysis were exactly the same as for the measurement on the 1.5 T MRI system. The total scan times for T_1 and B_1 mapping on the 3 T MRI system were 23.2 s and 27.2 s, respectively.

Image Reconstruction and Processing

The sampled k–space data were processed without applying any filter. The complex image data from each receive coil were combined using the sum–of–squares (SOS) algorithm proposed by Roemer et al.[64] Relative FA distribution $b_1(x, y)$ was assessed by determining the ratio of the effective FA $\alpha_{eff}(x, y)$ (Equation (1.12)) and the nominal FA α_{nom} (70°):

$$b_1(x, y) = \frac{\alpha_{eff}(x, y)}{\alpha_{nom}}. \tag{1.13}$$

The resulting b_1 map was smoothed by weighted polynomial fitting, and the matrix size was expanded by linear interpolation to 128×128. Finally, the T_1 map was determined by calculating T_1 values pixel–by–pixel using

$$T_1 = -\frac{T_R}{\ln\left(\frac{S_{\alpha_{1opt}}/\sin(b_1 \cdot \alpha_{1opt}) - S_{\alpha_{2opt}}/\sin(b_1 \cdot \alpha_{2opt})}{S_{\alpha_{1opt}}/\tan(b_1 \cdot \alpha_{1opt}) - S_{\alpha_{2opt}}/\tan(b_1 \cdot \alpha_{2opt})}\right)}, \tag{1.14}$$

where $S_{\alpha_{1opt}}$ and $S_{\alpha_{2opt}}$ represent the acquired signal amplitude of the T_1 mapping measurement using the optimum FAs.

Results

In Vitro Results

The unsmoothed b_1 map at its original image resolution of 64×64 and the FA–corrected T_1 map of the *in vitro* samples at 1.5 T are presented in Figure 1.3. The T_1 map was generated using the smoothed and resized b_1 map. For the 1.5 T and 3 T MRI systems, FA–corrected mean T_1 values are similar to reference T_1 values for each Gd–DOTA concentration (Table 1.1). Without FA correction, the mean T_1 values are constantly underestimated. The SDs of the uncorrected and FA–corrected T_1 values are similar.

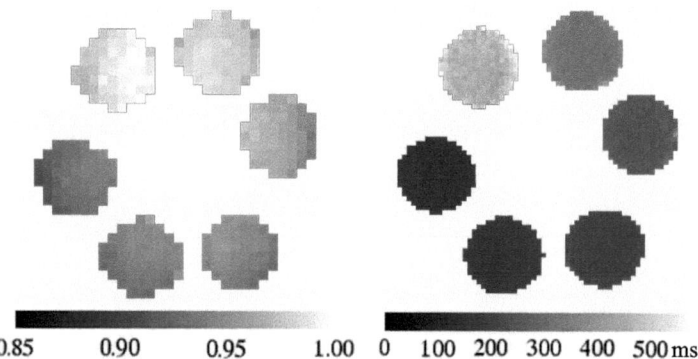

Figure 1.3: Grayscale–encoded b_1 map (left) and FA–corrected T_1 map (right) of six bottles containing different Gd–DOTA concentrations homogeneously mixed with a viscous glucose solution at 1.5 T. The b_1 values lower than 0.85 are in black. The different b_1 values for each bottle reflect the inhomogeneity of the transmitted B_1 field.

Table 1.1: *In vitro T_1 values for fast T_1 mapping and reference IR spectroscopy measurement*

Gd–DOTA concentration (μM)	1.5 T			3 T		
	T_1 mapping		Reference	T_1 mapping		Reference
	Uncorrected T_1 (ms)	Corrected T_1 (ms)	T_1 (ms)	Uncorrected T_1 (ms)	Corrected T_1 (ms)	T_1 (ms)
1200	121 (3)	138 (4)	134 (4)	127 (3)	154 (5)	143 (4)
1000	146 (4)	164 (4)	160 (4)	152 (4)	179 (5)	170 (4)
800	173 (5)	194 (5)	189 (4)	168 (6)	218 (6)	205 (4)
600	227 (6)	246 (6)	252 (4)	194 (10)	270 (9)	265 (4)
400	307 (7)	321 (6)	317 (4)	312 (9)	376 (13)	353 (4)
200	442 (12)	456 (13)	459 (4)	466 (14)	575 (16)	584 (4)

Data are expressed as median (standard deviation).

In Vivo Results

One representative smoothed b_1 map and the corresponding FA–corrected T_1 map of the abdomen at 1.5 T is shown in Figure 1.4. A continuous decrease in b_1 of 15 % is observed from right ($b_1 = 1.05$) to left ($b_1 = 0.9$) over the filled stomach (Figure 1.4a). The FA–corrected T_1 map displays a uniform distribution of the T_1 values within the ingested solution (Figure 1.4b). This uniformity is also reflected in the histogram plot in Figure 1.4c, demonstrating a normal distribution of the FA–corrected T_1 values. For uncorrected T_1 values, the distribution is positive–skewed due to the overestimation of the effective FAs. The FA–corrected mean T_1 value of the ingested glucose solution (199 ms (24 ms)) is similar to the *in vitro* reference T_1 value at 37°C (202 ms (4 ms)). For uncorrected T_1 values (188 ms (32 ms)) the mean is lower and the SD is higher compared to the FA–corrected T_1 values.

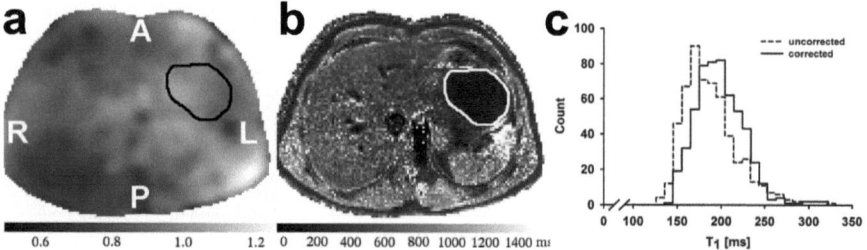

Figure 1.4: **(a)** Smoothed grayscale–encoded b_1 map (A, anterior; P, posterior; R, right; L, left) and **(b)** corresponding FA–corrected T_1 map of the abdomen with outlined stomach wall at 1.5 T. b_1 values lower than 0.5 are shown in black. **(c)** Histogram plot showing the distribution of the uncorrected and FA–corrected T_1 values of the ingested solution. The FA–corrected T_1 values are approximately normally distributed around the reference T_1 value of 202 ms.

In Figure 1.5 a representative smoothed b_1 map (a) and the corresponding FA–corrected T_1 map (b) at 3 T are displayed. The FA–corrected mean T_1 value of the ingested glucose solution (235 ms (27 ms)) is similar to the *in vitro* reference T_1 value at 37°C (216 ms (4 ms)). The uncorrected mean T_1 value (144 ms (28 ms)) of the glucose solution is much lower compared to the reference T_1 value.

Discussion

In this study a fast T_1 mapping technique based on the VFA approach was optimized achieving maximal T_1 accuracy over a T_1 range of 100 – 800 ms within a limited repetition T_R (9 ms) and acquisition time (2.3 s per T_1 map). Optimization methods were presented which maximize the SNR and ensure effective

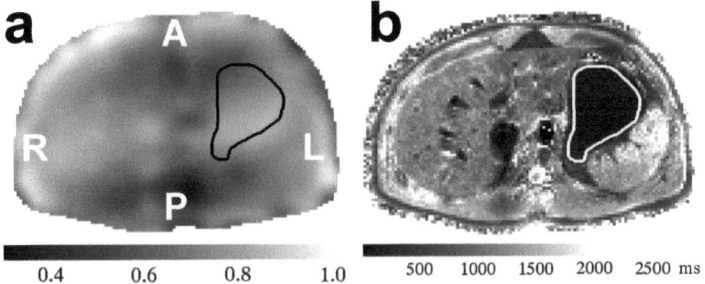

Figure 1.5: **(a)** *Smoothed grayscale–encoded b_1 map (A, anterior; P, posterior; R, right; L, left) and* **(b)** *corresponding FA–corrected T_1 map of the abdomen with outlined stomach wall at 3 T (5 min after meal ingestion). b_1 values lower than 0.3 are shown in black.*

RF and gradient spoiling, as well as a steady–state condition. Applying similar optimization schemes, a fast B_1 mapping technique was developed allowing the correction of spatial variations of the nominal FA over the excited FOV. Consecutive application of the T_1 and B_1 mapping sequence successfully generated high–precision T_1 maps of the human abdomen at a temporal resolution of 2.3 s per T_1 map.

The described optimization methods are effective for a T_1 range of 100 – 800 ms, and thus allow the assessment of changes in CA concentration for abdominal CE–MRI over a wide range of concentrations. The defined T_1 range corresponds to a typical T_1 range of (CE) abdominal organs at 1.5 T. T_1 values in the absence of CA are as follows: liver: $T_1 \sim 600$ ms;[65,66] kidney: $T_1 \sim 700$ ms;[65] spleen: $T_1 \sim 1000$ ms.[66] With the VFA technique there is always a trade–off between high T_1 accuracy (using more FAs) and high temporal resolution (using fewer FAs). As observed in a recent study,[62] multiple averaging of dual–angle T_1 maps provides higher accuracy per unit scan time than single–average, multiple FA data. However, to resolve the dynamics of orally or intravenously applied CAs, a high temporal resolution (approximately 2 s) of sequential T_1 maps is required.[56] Therefore, in this study a minimum number of two optimized FAs without signal averaging and a T_R of 9 ms were chosen for the T_1 mapping sequence, which resulted in a temporal resolution of 2.3 s per T_1 map at a voxel size of 2.8×2.8×15 mm³. To validate the proposed method *in vivo* in a human subject, an intragastric balloon was positioned inside a volunteer's stomach and filled with a homogeneously labeled viscous reference solution of known viscosity, temperature, and CA concentration. This approach presents the most reliable *in vivo* sequence validation, since the balloon prevents the reference solution from being diluted and mixed with an undefined volume of saliva and gastric juice, allowing the comparison of *in vitro* (at 37°C) and *in vivo* T_1 values.

The accuracy of the described T_1 mapping technique using the VFA approach decreases continuously as T_1 increases (Figure 1.6). This is confirmed by our *in vitro* results that show increased SDs for higher T_1 values. Furthermore, with higher T_1 values the acquired gradient–echo signal decreases (Equation (1.2)). Thus, for precise T_1 determination over a large T_1 range, it is of major importance to maintain the highest possible SNR. Multiple signal averaging is one way to increase SNR; however, this was omitted in this study in order to minimize the acquisition time per T_1 map. SNR was partly improved by using a large slice thickness (15 mm) and applying RF excitation pulses with high specific bandwidth, which resulted in improved slice excitation profiles.

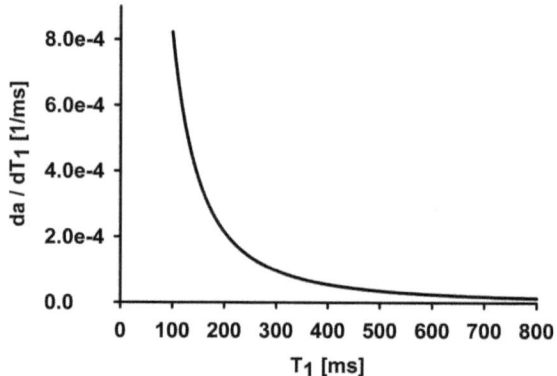

Figure 1.6: *Accuracy da/dT_1 of T_1 determination over a defined T_1 range of 100 – 800 ms. The accuracy decreases exponentially with increasing T_1 values.*

Image artifacts arising from imperfect spoiling and steady state impair accurate T_1 quantification. Such artifacts can be prevented by increasing the duration of the crusher gradients (increased dephasing of transverse spin components) and applying a sufficient number of dummy excitations. Since in this study the acquisition time was limited, optimization methods were developed ensuring effective gradient spoiling and steady–state condition. The effectiveness of the optimization methods was evaluated empirically. For gradient spoiling the amplitude of the crusher gradients was set to the maximum applicable value and the gradient duration was increased continuously until no ghosting artifacts were observed on the MR images. The steady–state condition as defined in Equation (1.8) was also derived empirically, and no blurring artifacts were visible in the MR images when the calculated number of dummy excitations was applied.

The dominant source of error (besides image artifacts) for T_1 quantification using the VFA technique are spatial variations of the nominal FA.[60] Through–plane variations arise from imperfect rectangular slice excitation profiles, and in–plane variations arise from B_1 field inhomogeneities. Effective slice excitation profiles were determined by computationally simulating the Bloch equations. For a sinc–Gaussian RF excitation pulse with a high specific bandwidth of 137 rad and a slice thickness of 15 mm, the slice profile errors are maximal for $\alpha_{2_{opt}} = 31°$ and $T_1 = 800$ ms. However, with these parameters, deviations from the nominal FA larger than 1° are restricted to only a small sublayer of 1–mm thickness on both sides of the excited slice. This indicates that slice profile errors have only a minor effect on the accuracy of T_1 quantification. B_1 field inhomogeneities, however, resulted in a significant underestimation of T_1 up to 10 % for 1.5 T, and even up to 30 % for 3 T. To date, only a few T_1 mapping studies of the human brain (using the VFA technique) have corrected B_1 field inhomogeneities.[60,67-69] These studies corrected B_1 field inhomogeneities by performing calibration measurements in homogeneous brain–like phantoms,[68,69] mapping the B_1 field using a segmented SE–EPI technique,[60] or applying three consecutive T_1–FFE sequences using FAs α, 2α, and 4α (i.e., three–point DESPOT1 technique).[67] Abdominal calibration measurements are not feasible due to respiratory and intestinal tissue motion. The application of the segmented SE–EPI sequence in the abdomen is critical because of its sensitivity to B_0 offsets. The three–point DESPOT1 technique has the disadvantage of requiring a fixed (not optimized) set of three FAs, which increases the temporal resolution per T_1 map by a factor of 3/2 compared to the VFA technique, which uses only two FAs.

In this study the B_1 field inhomogeneities in the abdomen were corrected by applying an optimized, fast B_1 mapping technique based on a spoiled gradient–echo sequence. For validation, a reference B_1 mapping technique (the double–angle method[70]) was performed *in vitro*. The present results are similar to those reported by Stollberger et al.,[70] but the temporal resolution of the reference method was much worse (410 s vs. 6.6 s per b_1 map). Because of the low spatial frequencies of the B_1 field distribution (especially at 1.5 T), a low image resolution of 64×64 was used and calculated b_1 maps were smoothed by appropriate weighted polynomial fitting to reduce the influence of noise. Gibb's ringing artifacts present in the low–resolution MRI datasets did not significantly affect the calculated b_1 maps. The long effective T_R of 120 ms makes this B_1 mapping technique sensitive to high–velocity pulsatile blood flow, as occurs in the aorta abdominalis and vena cava. For a slice thickness of 15 mm and peak flow velocities of $\sim$ 50 – 100 cm/s for the aorta abdominalis[71] and $\sim$ 50 cm/s for the vena cava,[72] the blood will completely move out of the excited slice before the next excitation pulse is applied. This causes signal variations in phase–encoding direction and thus ghosting artifacts of the aorta and vena cava in the image. Such ghosting artifacts can be avoided by measuring each k–space profile in the same heart phase using electrocardiogram (ECG) triggering. However, this would result in a considerably longer acquisition time. In this study, no ECG triggering was applied and phase–encoding was chosen in anterior–posterior direction of the transverse image slices to prevent the appearance of ghosting artifacts in the ROI.

Data processing was based on complex image data, which allowed the calculation of the actual MR signal amplitudes. B_1 field inhomogeneities were corrected by multiplying the optimum FAs of the T_1 mapping measurement by the relative FA distribution $b_1(x,y)$ derived from the B_1 mapping measurement (Equation (1.14)), assuming that $b_1(x,y)$ was identical for both measurements. This assumption holds true only in the case of a linear relationship between the nominal FA and the actual B_1 amplitude of the RF pulse. For slice–selective pulses, however, linearity represents only an approximation.[73] The deviation from the linear relationship depends on the applied RF pulse shape. Stollberger and Wach[70] showed that for sinc–shaped RF excitation pulses, linearity exists for FAs up to 140°, whereas for simple Gaussian–shaped pulses an essentially greater disproportionality will occur. In this study linearity was assumed, since for both T_1 and B_1 mapping measurements sinc–Gaussian–shaped RF excitation pulses with a high specific bandwidth of 137 rad and FAs = 70° were applied.

The fast T_1 mapping and B_1 mapping sequence was successfully implemented on two MRI systems of different magnetic field strength (1.5 T and 3 T). The results obtained at 3 T demonstrate the feasibility of this method for fast T_1 mapping at high magnetic field strengths. However, future applications will be restricted to the 1.5 T MRI system since an acquisition time of 5.8 s per T_1 map at 3 T is too long to assess CA dynamics in the abdomen. Therefore, the sequence parameters are optimized only for 1.5 T and not for 3 T. Nevertheless, the measured T_1 values of the ingested reference solution were similar to reference *in vitro* T_1 values at 37°C. The slightly higher mean T_1 value calculated 5 min after ingestion can be explained by dilution effects of the solution with saliva and gastric juice. The obtained *in vivo* results demonstrate the potential of the presented combined T_1 and B_1 mapping technique for fast, accurate abdominal T_1 quantification using MRI systems of different magnetic field strengths.

In conclusion, an optimized and combined T_1 and B_1 mapping technique was developed that is able to generate high–precision T_1 maps in the human abdomen at a temporal resolution of 2.3 s per T_1 map. By exploiting the one–

to–one relationship between T_1 and CA concentration, fast T_1 mapping allows the quantitative assessment of the distribution and dynamics of intravenously or orally applied CAs. The proposed fast T_1 mapping technique represents a promising noninvasive imaging method for a variety of clinical applications in abdominal CE–MRI, such as quantification of tissue perfusion and vascular permeability. Furthermore, this technique provides a valuable tool in gastrointestinal research for noninvasive and quantitative assessment of dilution, distribution, and mixing processes of labeled liquid solutions or pharmacological substances in the gastrointestinal tract.[74]

Chapter 2

Fast T_1 Mapping for the Noninvasive Quantification of Gastric Secretion

Introduction

Quantification of gastric secretion is of key importance for the understanding and effective management of many gastrointestinal diseases such as peptic ulcer and gastroesophageal reflux.[23] Aspiration and *in vivo* intragastric titration represent the two most commonly applied techniques to measure intragastric acid secretion.[75,76] These methods are invasive, unphysiological and may stimulate gastric secretion by themselves. Moreover, both are applicable only with experimental, liquid test meals and allow only a global determination of gastric secretion although it is known that mixing of gastric contents is poor and local distribution of gastric secretion is often inhomogeneous (e.g. layering).[23,77] Alternative approaches for the assessment of gastric secretion, noninvasive magnetic resonance imaging (MRI) techniques based on gastric volume measurements[78] and T_2 mapping of viscous model meals using *in vitro* calibration of $(T_2)^{-1}$ against polysaccharide concentration,[79,80] have been presented. However, changes in gastric volume do not only reflect dynamics of gastric secretion but a combination of secretion and emptying and the T_2 mapping technique can only be applied to polysaccharide based viscous meals.

Recently, a fast and optimized T_1 mapping technique for the quantification of dilution and mixing processes of orally applied gadolinium (Gd) based paramagnetic contrast agents (CAs) in human abdominal MRI was developed.[81] This technique can be extended to the quantitative assessment of gastric secretion volume: by homogeneously labeling a meal with a CA and dynamically detecting the changes in CA concentration (C_{Gd}) by measuring changes in the relaxation time T_1, the volume of gastric secretory products can be determined. Hereby, the interrelationship of the relaxation time T_1 and C_{Gd} basically is given

by:
$$\frac{1}{T_1} = \frac{1}{T_{10}} + r_1 \cdot C_{Gd}, \qquad (2.1)$$

where r_1 is the relaxivity of the CA and T_{10} the relaxation time of the meal in the absence of a CA.[43] However, it has to be regarded that for a labeled meal which is continuously diluted and mixed with gastric secretion the relaxivity r_1 and T_{10} change depending on various parameters such as temperature, pH and macromolecular content of the diluted and mixed meal.[82]

The aim of the present study was to evaluate the noninvasive quantification of gastric secretion volume after administration of a paramagnetic labeled viscous glucose solution (test meal) by measuring T_1. An *in vitro* calibration curve describing the interrelationship between T_1 and C_{Gd} as well as the dependency of r_1 and T_{10} on the macromolecular concentration was derived. To analyze the accuracy and applicability of the $T_1 - C_{Gd}$ calibration curve *in vivo*, T_1 mapping of four glucose solutions labeled with different C_{Gd} was performed in five healthy volunteers of different body mass index (BMI). Furthermore, *ex vivo* T_1 and pH measurements of gastric (GJ) and duodenal juice (DJ) were performed in order to estimate the relaxation properties and the intersubject variability of the secreted gastroduodenal juice and also the dependency of T_1 on pH.

Materials and Methods

The optimized fast T_1 mapping technique used in this study is based on the variable flip angle approach.[58,59] Spatial variations of the nominal flip angle due to B_1 inhomogeneities are corrected by performing an additional optimized fast B_1 mapping technique based on a gradient echo sequence with different alternating repetition times.[61]

The methodological part is subdivided into three sections. In the first section, *ex vivo* measurements and analysis of T_1 and pH of gastric juice (GJ) and duodenal juice (DJ) from fourteen healthy volunteers are described. In the second section, the *in vitro* interrelationship between T_1 and C_{Gd} of a labeled and diluted viscous glucose solution and the dependency of r_1 and T_{10} on the macromolecular concentration are derived. Finally, in the third section, the *in vivo* validation of the $T_1 - C_{Gd}$ calibration curve and the measurement technique in five healthy volunteers is presented.

All MRI measurements were performed on a 1.5 T whole–body MRI system (1.5 T Achieva, Philips Medical Systems, Best, The Netherlands) using an abdominal, four–channel phased–array receive coil.

Ex vivo T_1 and pH Measurements of Gastric and Duodenal Juice

GJ and DJ of fourteen healthy volunteers (seven women and seven men; mean age = 27 years (range = 21 – 33 years); mean body mass index (BMI) = 23.2 kg/m^2 (range = 20.3 – 28.0 kg/m^2)) with no history of gastrointestinal disease were analyzed. Written informed consent was obtained from each volunteer and the examination protocol was approved by the local ethics committee. Measurements were performed in the morning of three different study days after an overnight fast. One volunteer could only be investigated on two study days. For each volunteer, 15 ml of GJ and DJ was aspirated using a nasoduodenal tube[83] and obtained juice was filled in plastic syringes. In addition, three syringes filled with 15 ml of a 0.1 N (pH = 1) HCl solution were prepared on three different study days. T_1 measurements of the solutions at room temperature were performed by applying a standard inversion recovery (IR) spectroscopy sequence (T_R = 15 s, six inversion delay times equally spaced by 10 ms and appropriately chosen to detect the zero–crossing of the signal amplitude in the spectrum). T_1 was determined by detecting the zero–crossing of the signal amplitude at time point τ and using the interrelationship $T_1 = \tau/ln2$. pH was measured by pH indicator paper (Universalindikator, Merck (Switzerland) AG, CH–8953 Dietikon, Switzerland).

In vitro $T_1 - C_{Gd}$ Calibration Curve

Four homogeneous solutions were obtained by serial dilution of a viscous glucose solution (10 % glucose homogeneously mixed with 1 % locust bean gum (LBG) powder) with appropriate volumes of a 0.1 N HCl solution designed to mimic mixing of a viscous meal with gastric secretion across a wide range of dilutions. The increased viscosity of the glucose solution reduced potential flow artifacts on the MR images. *(Initial) solution 1*: no dilution; *solution 2*: solution 1 diluted with HCl at a ratio of 5:1; *solution 3*: solution 1 diluted with HCl at a ratio of 1:1; *solution 4*: solution 1 diluted with HCl at a ratio of 1:5. For each of the four solutions six samples (100 ml each) containing different Gd–DOTA (DOTAREM®, Laboratoire Guerbet, France) concentrations (1200 μM, 1000 μM, 800 μM, 600 μM, 400 μM and 200 μM) at a temperature of 37°C were prepared in small plastic bottles.

A total of four MRI measurements were performed, one measurement for all six samples of each solution. The six plastic bottles were simultaneously placed in the isocenter of the MRI system and one image slice aligned perpendicular to the long axes of the bottles was acquired. For T_1 determination an optimized and combined T_1 and B_1 mapping technique, which is described in detail in Chapter 1, was applied. Sequence parameters were as follows: FOV = 360 mm, slice thickness = 15 mm; T_1 *mapping*: flip angles = 5° and 31°, number of dummy excitations = 29 and 21, scan matrix = 128×128, T_R/T_E = 9/3.6 ms, scan time = 2.3 s per T_1 map; B_1 *mapping*: flip angle = 70°, number of dummy excitations = 6, scan matrix = 64×64, T_{R_1} = 20 ms, T_{R_2} = 100 ms, T_E = 3.6 ms, scan time = 6.6 s per B_1 map.

T_1 values (mean (standard deviation)) were calculated within a region of interest (ROI) of 6×6 pixels for each bottle. Parameters T_{10} and r_1 were fitted to the model as described by Equation (2.1) using a two–parameter nonlinear least squares fit based on the Levenberg–Marquardt (LM) algorithm.[84] This results in a set of two parameters T_{10} and r_1 for each solution. The dependency of T_{10} on the relative macromolecular concentration C_M (defined as the ratio of macromolecules in the diluted solution to macromolecules in *initial solution 1* and expressed as percentage) and r_1 on C_M was determined using a nonlinear least squares fit based on the LM algorithm. In order to determine the interrelationship between T_1 and C_{Gd} for the continuously diluted labeled *solution 1*, the obtained dependency of T_{10} on C_M as well as r_1 on C_M was substituted in Equation (2.1) using

$$C_M = \frac{100}{1200} \cdot C_{Gd}, \qquad (2.2)$$

where C_M is expressed in % and C_{Gd} in μM.

In vivo Validation of T_1 − C_{Gd} Calibration Curve

The T_1 − C_{Gd} calibration curve was validated *in vivo* in five healthy volunteers (one woman and four men; mean age = 30 years (range = 24 − 37 years)) of different BMI (mean BMI = 23.9 kg/m² (range = 20.3 − 28.1 kg/m²)). Thereby, the effect of different object load (different B_1 distribution and noise) on the accuracy of T_1 determination could be analyzed.

Written informed consent was obtained from each volunteer and the protocol was approved by the local ethics committee. After six hours fasting, volunteers arrived at the MR center and were asked to drink 100 ml of apple juice to induce post–prandial antral contraction patterns (i.e. non–occlusive luminal contractions) during measurements.[85] After drinking, volunteers were intubated nasally with a small balloon attached to the end of a nasogastric tube. The balloon was positioned within the proximal part of the stomach. For subsequent imaging, volunteers were placed in left decubitus body position in the MRI system. Four test meals at a temperature of 37°C were prepared by diluting the labeled *solution 1* (10 % glucose, 1 % LBG, 1200 μM Gd–DOTA) with appropriate volumes of a 0.1 N HCl solution (*solution A*: no dilution; *solution B*: *solution 1* diluted with HCl at a ratio of 2:1; *solution C*: *solution 1* diluted with HCl at a ratio of 1:2; *solution D: solution 1* diluted with HCl at a ratio of 1:5).

Examination procedure was started by filling the balloon with 300 ml of *solution A* using a plastic syringe. Then, four oblique transverse image slices aligned perpendicular to the proximal stomach axis were acquired within one breath–hold for both the T_1 mapping (total scan time = 9.2 s) and subsequent B_1 mapping (total scan time = 24.6 s) sequence. Scan parameters were the same as described in the previous section. When combined T_1 and B_1 mapping measurements were finished *solution A* was retrieved by aspiration. This "filling–measurement–emptying" procedure was repeated consecutively for *solution B, C* and *D* while volunteers remained in left decubitus body position.

For T_1 determination only 65 of total 80 T_1 maps were analyzed. On 15 image slices pronounced flow artifacts were observed which would artificially increase the error in T_1 determination.

Statistical Analysis

Statistical analysis was performed using SPSS® for Windows Release 13 Software (SPSS Inc., Chicago, IL, USA). Distribution of *ex vivo* T_1 and pH values of GJ and DJ were analyzed using a one–sample Kolmogorov–Smirnov test. A paired–samples t–test was applied to analyze differences in T_1 and a Wilcoxon signed–rank test for differences in pH of GJ and DJ. The correlation between T_1 and pH was analyzed using two–tailed Pearson correlation. For the *in vivo* experiments, distribution of T_1 values was analyzed by applying a one–sample Kolmogorov–Smirnov test. All data are expressed and presented as mean (standard deviation). A P–value < 0.05 was considered statistically significant.

Results

Ex vivo T_1 and pH Measurements of Gastric and Duodenal Juice

Results show that T_1 of GJ was significantly longer than T_1 of DJ (GJ: 2939 ms (114 ms) vs. DJ: 2858 ms (110 ms), P < 0.001; Figure 2.1); however, the difference in T_1 was small and much less compared to changes observed for diluted viscous glucose solutions (see section below). As expected, pH values were significantly lower in GJ than DJ (GJ: 1.4 (0.7) vs. DJ: 7.1 (0.9), P < 0.001). The statistical analysis indicates a trend ($P = 0.07$) for a correlation between T_1 and pH ($r = -0.78$). Mean T_1 value of the HCl solution was 2760 ms.

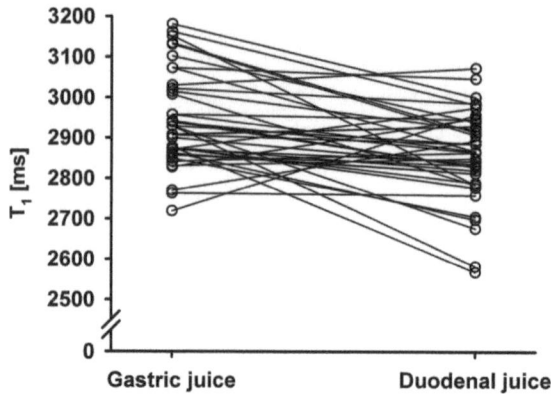

Figure 2.1: T_1 values of gastric and duodenal juice measured ex vivo at room temperature. Corresponding data points for each volunteer and measurement are connected by a straight line. The slopes of the connecting lines indicate a longer T_1 of gastric juice compared to duodenal juice.

In vitro $T_1 - C_{Gd}$ Calibration Curve

One representative T_1 map for *solution 1* is shown in Figure 2.2a. For each of the four solutions, measured T_1 values together with nonlinear least squares fit are plotted in Figure 2.2b. A linear relationship between r_1 and C_M (Figure 2.3a) and between T_{10} and C_M (Figure 2.3b) was found. The resulting calibration curve of T_1 vs. Gd–DOTA concentration C_{Gd} for the viscous glucose solution (10 % glucose, 1 % LBG, 1200 μM Gd–DOTA) that was continuously diluted with 0.1 N HCl at 37°C is presented in Figure 2.4. Corresponding equation was calculated as:

$$\frac{1}{T_1} = \frac{1}{a \cdot C_{Gd} + b} + (c \cdot C_{Gd} + d) \cdot C_{Gd}, \tag{2.3}$$

with the empirically derived parameters $a = 4.84 \cdot 10^{-1}\ ms \cdot \mu M^{-1}$; $b = 2.20 \cdot 10^3\ ms$; $c = 7.46 \cdot 10^{-10}\ ms^{-1} \cdot \mu M^{-2}$ and $d = 2.54 \cdot 10^{-6}\ ms^{-1} \cdot \mu M^{-1}$. T_1 and C_{Gd} are expressed in ms and μM, respectively.

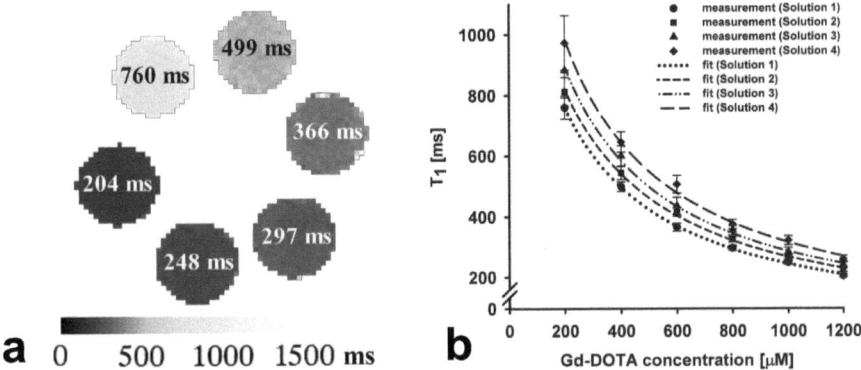

Figure 2.2: **(a)** *Grayscale-encoded T_1 map of six bottles containing different Gd–DOTA concentrations homogeneously mixed in solution 1 (10 % glucose, 1 % LBG) at 37°C. Mean T_1 values derived from a region of interest of 6×6 pixels for each bottle are presented.* **(b)** *Measured T_1 values (symbols) and nonlinear least squares fits according to Equation (2.1) (dashed lines) of four diluted glucose solutions homogeneously mixed with different Gd–DOTA concentrations at 37°C. Data points are plotted as mean ± standard deviation.*

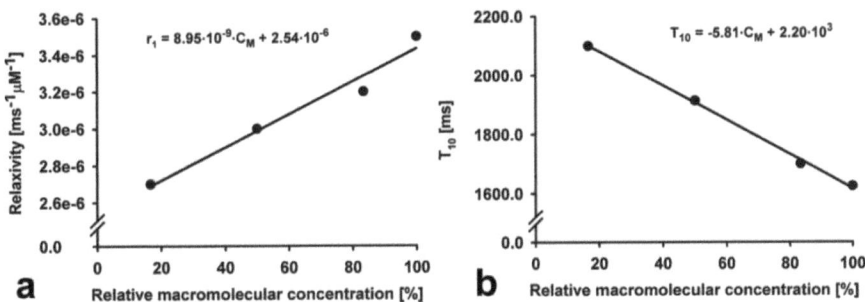

Figure 2.3: Relaxivity r_1 **(a)** *and T_{10}* **(b)** *as a function of the relative macromolecular concentration C_M (defined as the ratio of macromolecules in the diluted solution to macromolecules in initial solution 1 and expressed as percentage). The interrelationship between r_1 and C_M as well as T_{10} and C_M is well described by linear regression (***(a)*** $R^2 = 0.965$;* **(b)** *$R^2 = 0.998$)*

In vivo Validation of $T_1 - C_{Gd}$ Calibration Curve

In vivo mean (SD) T_1 values for each viscous glucose solution are presented in Table 2.1. Results show no dependency of T_1 on the BMI of the volunteers. A series of representative abdominal T_1 maps of one healthy volunteer is shown in Figure 2.5a. The excellent agreement between *in vivo* mean T_1 values of all volunteers and the reference *in vitro* calibration curve is illustrated in Figure 2.5b. A more detailed representation of the distribution of the measured *in*

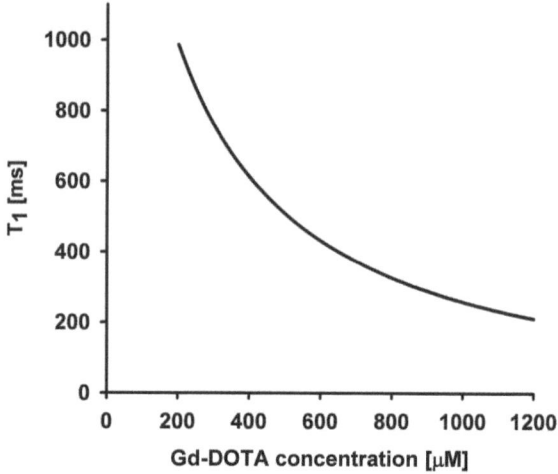

Figure 2.4: T_1 as a function of the Gd–DOTA concentration (C_{Gd}) for labeled solution 1 (10 % glucose, 1 % LBG, 1200 µM Gd–DOTA) that was continuously diluted with 0.1 N HCl at 37°C (Equation (2.3)).

vivo T_1 values is depicted in Figure 2.5c. Results demonstrate that T_1 values are normally distributed ($P < 0.05$) around the reference in vitro T_1 value for each of the four glucose solutions. The standard deviation (SD) of the T_1 distribution increases for decreasing C_{Gd} (SD, 33 ms for 1200 µM vs. 55 ms for 800 µM vs. 109 ms for 400 µM vs. 210 ms for 200 µM). Despite decreasing accuracy with increasing dilution, the full width at half maximum (FWHM) of the T_1 distributions did not overlap and, thus, the different dilutions could be distinguished using the optimized fast T_1 mapping technique (Figure 2.5c).

Table 2.1: *In vivo T_1 values*

	Solution A T_1 (ms)	Solution B T_1 (ms)	Solution C T_1 (ms)	Solution D T_1 (ms)
Volunteer 1 (BMI = 20.3 kg/m²)	198 (34)	348 (52)	636 (114)	991 (190)
Volunteer 2 (BMI = 22.3 kg/m²)	223 (29)	331 (44)	607 (112)	1008 (220)
Volunteer 3 (BMI = 23.8 kg/m²)	204 (27)	335 (47)	582 (102)	960 (191)
Volunteer 4 (BMI = 24.8 kg/m²)	215 (30)	290 (54)	578 (109)	955 (213)
Volunteer 5 (BMI = 28.1 kg/m²)	186 (31)	327 (54)	553 (98)	1019 (225)
Mean of all volunteers	207 (33)	324 (55)	587 (109)	986 (210)

Data are expressed as mean (standard deviation).

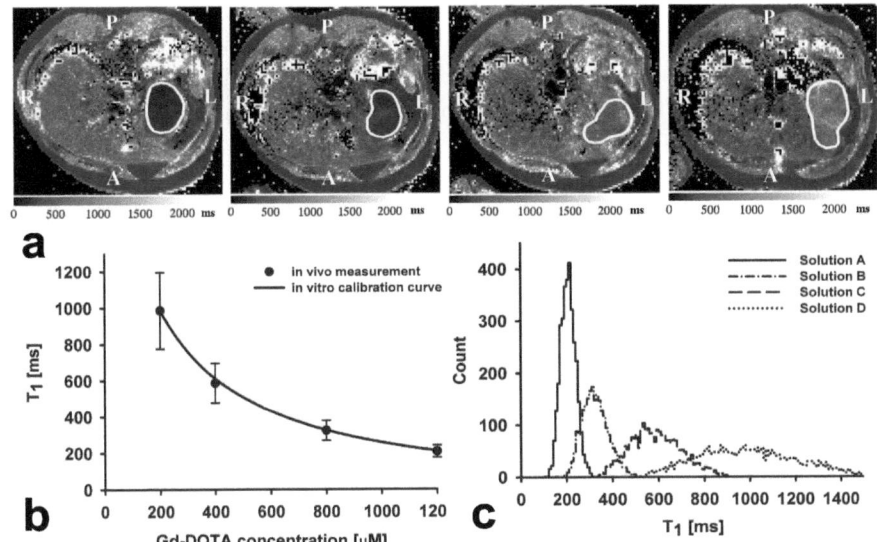

Figure 2.5: **(a)** *Series of four representative abdominal T_1 maps (A, anterior; P, posterior; L, left; R, right) of one healthy volunteer for solutions A, B, C and D (from left to right). The stomach wall is outlined on each T_1 map.* **(b)** *Mean ± SD T_1 values of all volunteers (circles) together with in vitro calibration curve (line) as derived from Equation (2.3). The calibration curve is accurately sampled by the in vivo mean T_1 values.* **(c)** *Mean T_1 distributions of all volunteers for each of the four labeled glucose solutions (reference in vitro T_1 values: solution A, 211 ms; solution B, 327 ms; solution C, 612 ms; solution D, 986 ms). Areas under the curves are not identical.*

Discussion

In this study, the noninvasive quantification of the dilution of a labeled viscous glucose solution with HCl (used as surrogate for gastric secretion) by measuring the relaxation time T_1 was successfully evaluated. Using the interrelationship between T_1 and Gd–DOTA concentration (C_{Gd}), dynamic T_1 mapping allows the determination of the change in C_{Gd} and thus the distribution of gastric secretion volume. So far, there are only two published studies, also applying MRI, demonstrating a noninvasive technique for the determination of gastric secretion volume.[80,86] In these studies, gastric secretion volume was assessed by detecting changes in meal viscosity using *in vitro* calibration of transverse relaxation rate $(T_2)^{-1}$ against polysaccharide concentration of viscous model meals. Thus, this T_2 mapping technique is limited to polysaccharide based viscous meals whereas the method here presented can be applied in combination with any macronutrient liquid meal labeled with a paramagnetic contrast agent.

T_1 and pH values of GJ measured *ex vivo* were close to those of a 0.1 N

HCl solution. This is an expected finding since hydrochloric acid represents the main component of gastric juice[87] and HCl and GJ have a similar aqueous composition. Due to the limited sensitivity of the fast T_1 mapping technique for long T_1 values as further described below and in the paper of Treier et al.,[81] small differences in T_1 between GJ and HCl cannot be resolved using the here presented technique. Therefore, serial dilutions with HCl were used to simulate the post–prandial dilution of a viscous glucose solution (test meal) with gastric secretion for the *in vitro* $T_1 - C_{Gd}$ calibration curve. However, in order to detect possible differences between long T_1 values as is the case for GJ and DJ, a standard IR spectroscopy sequence with a prolonged acquisition time and thus increased sensitivity was applied for T_1 measurements *ex vivo*. Results showed a slightly but significantly longer T_1 of GJ compared to DJ. This difference in T_1 was not well explained by the lower pH value of GJ. A more detailed analysis of GJ and DJ concerning macromolecular composition would be necessary but was beyond the scope of this study.

The relaxivity r_1 of Gd–DOTA within the viscous glucose test meal and the relaxation time T_{10} depend in a complex way on the magnetic field strength, temperature, pH and macromolecular content of the meal.[82] The influence of field strength can be ignored since all MRI measurements were performed at 1.5 T. Results in a previous study[81] demonstrated that temperature has a major effect on T_1 of the applied labeled glucose solution at 1.5 T with longer T_1 values at body temperature of 37°C (202 ms (4 ms) compared to T_1 values at room temperature (134 ms (4 ms)). Therefore, to avoid temperature dependent changes in T_1, all *in vitro* measurements were performed at 37°C. The influence of pH on T_1 can be neglected as demonstrated in the results of the *ex vivo* experiments of GJ and DJ. However, macromolecular concentration of the meal was shown to cause significant changes in r_1 and T_{10}. A linear relationship between r_1 and macromolecular concentration and between T_{10} and macromolecular concentration was found *in vitro*. This is in agreement to theoretical models described in literature.[88,89] An increase in macromolecular content causes a reduction of the correlation times of the Gd–DOTA chelate or the water molecule, or both, leading to an increased rate of spin interaction.

The presented technique for the noninvasive quantification of gastric secretion volume was successfully validated in five healthy volunteers of different BMI. The *in vitro* $T_1 - C_{Gd}$ calibration curve is accurately sampled by the four data points obtained from *in vivo* measurements. For *in vivo* validation, a small balloon was positioned in the proximal stomach and filled with different labeled viscous glucose solutions of known T_1. This represents the most reliable approach for *in vivo* validation since the balloon prevents the solution from being diluted with undefined volumes of ingested apple juice, saliva and GJ assuring identical meal composition as for the *in vitro* reference measurements. For each volunteer, T_1 mapping measurements of four different solutions were performed according to the "filling–measurement–emptying" procedure described in the materials and methods section. Using a syringe for emptying it is not possible to retrieve the total volume of the ingested solution. Potentially, this could result in a slightly different meal composition and thus T_1 value; however, as confirmed in a separate *in vitro* experiment, the error in T_1 induced by this procedure was assumed less than 1 %.

T_1 values for each solution were normally distributed around the reference *in vitro* T_1 value. Deviations from the normal T_1 distribution would indicate a inhomogeneous dilution of the meal. Flow artifacts within the stomach secondary to gastric contractions could also result in deviations from the normal distribution (broadening of the T_1 distribution, displacement of the peak). Thereby, flow is less pronounced in viscous solutions (e.g. as in a normal meal) and in the post–prandial proximal stomach (i.e. usual location for measurement of gastric secretion). Only 15/80 slices could not be analyzed due to pronounced flow artifacts. Motion artifacts were not observed on the images of both T_1 and B_1 mapping scan although breath–hold for B_1 mapping was relatively long. This was possibly due to the low image resolution for B_1 mapping. *In vivo* results further demonstrated an increase in the standard deviation of the measured T_1 distribution for low contrast agent concentrations. In order to analyze the limitation of the achievable T_1 or C_{Gd} resolution using the optimized fast T_1 mapping technique, a theoretical model was derived estimating the width of the T_1 distribution as a function of the nominal T_1 value. The error propagation function for the basic equation of the T_1 mapping technique (Equation (A2.1)) was estimated using a three–parameter (a, b, c) nonlinear least squares fit of the form $a \cdot \sqrt{\left(\frac{\partial f}{\partial S_1}\right)^2 \cdot b + \left(\frac{\partial f}{\partial S_2}\right)^2 \cdot c}$ (see equation (A2.3)) to the measured standard deviations for *solution A, B, C* and *D* (full derivation in the Appendix). The increase in the standard deviation of the T_1 distribution as a function of the nominal T_1 value is observed to be close to linearity (Figure 2.6). Based on this theoretical model, using the optimized fast T_1 mapping technique, it will not be possible, for example, to differentiate between gastric and duodenal secretion. Rather, for accurate analysis of gastric secretion volume, appropriate CA concentrations must be applied to achieve high accuracy in T_1 determination (i.e. short nominal T_1 values).

In conclusion, a technique for the quantitative and noninvasive assessment of intragastric secretion volume based on the interrelationship between T_1 and CA concentration of a labeled viscous glucose solution was developed *in vitro* and validated in healthy volunteers. This technique allows the noninvasive and

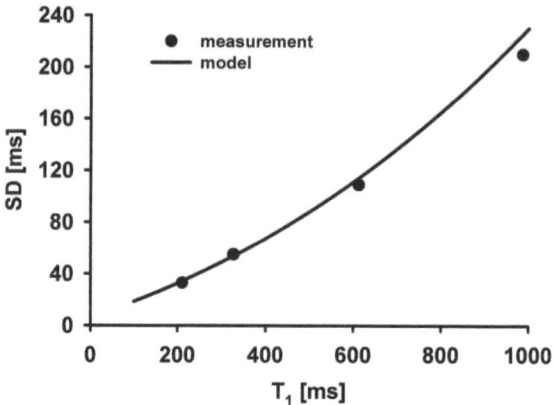

Figure 2.6: Measured standard deviations (circles) of the T_1 distribution for solution A, B, C and D as well as theoretical model (line) based on Equation (A2.3).

three–dimensional quantification and visualization of gastric secretory products distribution in the human stomach and therefore provides a valuable tool for the evaluation of the efficacy of drugs to stimulate or inhibit gastric secretion. In an ongoing study, the effect of pentagastrin stimulated gastric secretion on gastric volume responses, emptying and intragastric distribution in healthy volunteers is investigated.[90] A further field of application of this technique presents the *in vivo* quantitative determination of gastric and intestinal mixing and distribution of orally administered paramagnetic labeled substances or dissolving drug delivery systems[39,40] in pharmacological studies of controlled release by dynamic T_1 mapping.

Appendix

The calculation of T_1 using the variable flip angle approach is a function f of the measured gradient echo signals S_1 and S_2, corresponding flip angles α_1 and α_2 and repetition time T_R:

$$T_1 = -\frac{T_R}{\ln\left(\frac{S_1/\sin\alpha_1 - S_2/\sin\alpha_2}{S_1/\tan\alpha_1 - S_2/\tan\alpha_2}\right)} \equiv f(S_1, S_2). \tag{A2.1}$$

Following the Gaussian law of error propagation (for small errors), the variance σ_f^2 of calculated T_1 is given by:[91]

$$\sigma_f^2 = \left(\frac{\partial f}{\partial S_1}\right)^2 \cdot \sigma_{\bar{S_1}}^2 + \left(\frac{\partial f}{\partial S_2}\right)^2 \cdot \sigma_{\bar{S_2}}^2, \tag{A2.2}$$

where $\sigma_{\bar{S_1}}^2$ and $\sigma_{\bar{S_2}}^2$ are the variances of the mean signals S_1 and S_2.

Taking the square root of Equation (A2.2) results in an expression for the error (as described by the standard deviation SD) of T_1 determination

$$SD(T_1) = \sqrt{\left(\frac{\partial f}{\partial S_1}\right)^2 \cdot \sigma_{\bar{S_1}}^2 + \left(\frac{\partial f}{\partial S_2}\right)^2 \cdot \sigma_{\bar{S_2}}^2} \tag{A2.3}$$

with

$$\frac{\partial f}{\partial S_1} = \frac{C \cdot T_R}{\left[\ln\left(\frac{-C \cdot exp\left(-\frac{T_R}{T_1}\right)}{D_1 \cdot \cos\alpha_2 - D_2 \cdot \cos\alpha_1}\right)\right]^2} \cdot \frac{D_1^2 \cdot D_2}{\left(-C \cdot exp\left(-\frac{T_R}{T_1}\right) \cdot \sin\alpha_1\right) \cdot \left[\left(1 - exp\left(-\frac{T_R}{T_1}\right)\right) \cdot (D_1 \cdot \cos\alpha_2 - D_2 \cdot \cos\alpha_1)\right]} \tag{A2.4}$$

and

$$\frac{\partial f}{\partial S_2} = -\frac{C \cdot T_R}{\left[\ln\left(\frac{-C \cdot exp\left(-\frac{T_R}{T_1}\right)}{D_1 \cdot \cos\alpha_2 - D_2 \cdot \cos\alpha_1}\right)\right]^2} \cdot \frac{D_1 \cdot D_2^2}{\left(-C \cdot exp\left(-\frac{T_R}{T_1}\right) \cdot \sin\alpha_2\right) \cdot \left[\left(1 - exp\left(-\frac{T_R}{T_1}\right)\right) \cdot (D_1 \cdot \cos\alpha_2 - D_2 \cdot \cos\alpha_1)\right]} \tag{A2.5}$$

where $C = \cos\alpha_1 - \cos\alpha_2$; $D_1 = 1 - exp\left(-\frac{T_R}{T_1}\right) \cdot \cos\alpha_1$; $D_2 = 1 - exp\left(-\frac{T_R}{T_1}\right) \cdot \cos\alpha_2$.

Chapter 3

Gastric Motor Function and Emptying in the Right Decubitus and Seated Body Position as Assessed by Magnetic Resonance Imaging

Introduction

Magnetic resonance imaging (MRI) has been established as a valuable technique in human gastrointestinal (GI) research for analyzing gastric function.[30,31,80,92] In comparison with radionuclide and ultrasound imaging methods, MRI offers improved spatial and temporal image resolution, and thus is ideal for noninvasive and reliable assessment of GI physiology.[93] It will therefore play an important role in GI research and clinical diagnosis in the future.[86,94]

The most common body position during and following meal ingestion is sitting; however, the horizontally aligned whole–body architecture of modern high–field MRI systems restricts measurements of organ function to the lying body position. Several studies using γ–scintigraphy,[34–38] intraluminal manometry,[35] and MRI[95] have shown that posture influences gastric function, and thus may present a limitation for gastric MRI. In those studies, however, gastric emptying was analyzed only at discrete time points, and peristaltic motility was determined as antropyloroduodenal pressure events. Differences in the gastric emptying rate and the number of antropyloric contraction waves were detected between the lying and seated positions. Because of the very different measurement principles involved (i.e., radioactive decay vs. hydrostatic pressure vs. MR), scintigraphy and manometry data are not directly comparable with data derived using MRI. To evaluate gastric MRI for its use in clinical research and diagnosis of GI pathophysiology, one must simultaneously investigate gastric emptying, stomach volume, intragastric meal distribution and gastric peristalsis in the lying position and detect differences compared to the upright, seated body position (SP).

The aim of this study was to determine the effect of the right decubitus lying body position (RP) on relevant parameters of human gastric motor function in healthy volunteers as assessed by MRI. In this position, intragastric meal distribution is most similar to that in SP because the distal stomach is always filled with gastric contents, and intragastric air is confined to the proximal stomach. Imaging protocols were developed for two MRI systems that differed in architecture, thus allowing intraindividual comparisons of stomach and intragastric air volume, intragastric meal distribution, gastric emptying, and gastric peristalsis between RP and SP.

Materials and Methods

Subjects

Ten healthy volunteers (four women and six men, mean age = 25.6 years, range = 20 – 34 years) participated in this study. None of the subjects had a history of gastrointestinal disease or took any medication that affected gastric function. Written informed consent was obtained from each volunteer, and the study protocol was approved by the local ethics committee.

Study Protocol

Measurements were performed on two different study days in RP and SP. Volunteers were examined in the afternoon after they fasted for six hours. Three volunteers had to be examined in the morning in SP due to limited MRI scanner availability. Six volunteers started in RP and the other four began in SP. On both study days the same low fat solid/liquid test meal was administered. The solid phase consisted of 150 g cooked pasta (214.5 kcal; protein: 9 g; carbohydrates: 39 g; fat: 2.25 g) and the liquid phase of 150 ml of a nutrient drink (Ensure® Plus Drink, 225 kcal, protein: 9.375 g; carbohydrates: 30.3 g; fat: 7.38 g) labeled with 0.5 mM of the paramagnetic contrast agent Gd–DOTA (DOTAREM®, Laboratoire Guerbet, France). The solid meal was always administered prior to the liquid meal.

Measurements in RP were performed using the 1.5 T whole–body MRI system (1.5 T Intera, Philips Medical Systems, Best, The Netherlands) as shown in Figure 3.1a. Volunteers ingested the solid meal in SP in the scanner room. Immediately afterwards they were placed in RP inside the whole–body MRI system, and they ingested the liquid meal. Measurements in SP were performed using the 0.5 T open–configuration MRI system (Signa SP/i 0.5 T, GE Medical Systems, Milwaukee, WI, USA) as shown in Figure 3.1b. Volunteers ingested both solid and liquid meals in SP inside the open–configuration MRI system.

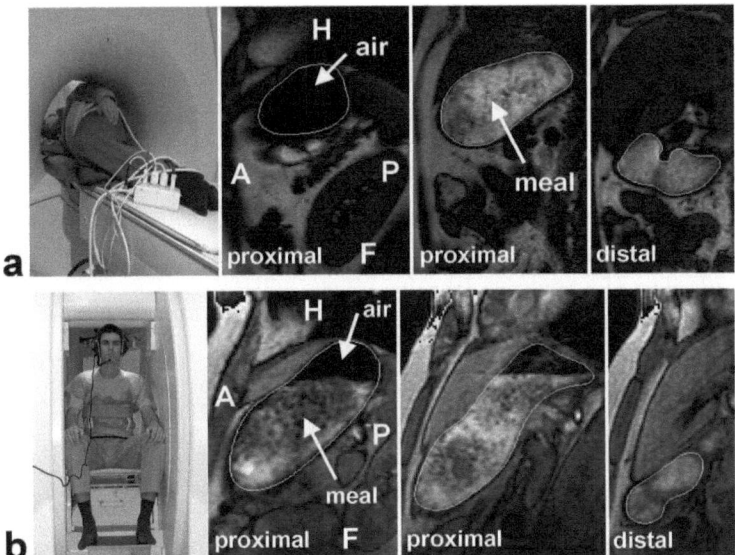

Figure 3.1: **(a)** Left: Volunteer in RP inside the 1.5 T whole–body MRI system. Rectangular surface coils were fixed around the abdomen. Right: Three sagittal MR image slices (two proximal and one distal) of a volume scan in RP with outlined stomach wall. Air and meal are indicated in the MR images by arrows (F, feet; H, head; A, anterior; P, posterior). **(b)** Left: Volunteer sitting inside the 0.5 T open–configuration MRI system. A send/receive coil was fixed around the abdomen. Right: Three sagittal MR image slices (two proximal and one distal) of a volume scan in SP with outlined stomach wall. Air and meal are indicated in the MR images by arrows (F, feet; H, head; A, anterior; P, posterior).

A scan consisting of 20 sagittal image slices covering the complete gastric region (*volume scan*) was performed in the fasted state ($t = -15$ min) and after solid meal intake ($t = -4$ min). After ingestion of the liquid meal at time $t = 0$ min, *volume scans* followed by a dynamic scan sequence (*motility scan*) were performed at 5 min intervals until time $t = 30$ min, and thereafter at 10 min intervals until the end of the study period ($t = 90$ min). Based on the preceding volume data, the image stack of the *motility scan* was positioned along the distal stomach axis in order to detect the propagating gastric contraction waves.

On the 1.5 T MRI system, steady–state free precession (SSFP) sequences were used for imaging. For the *motility scan* the parallel imaging method sensitivity encoding (SENSE)[32] was applied to increase the image acquisition rate. The imaging parameters for this system were as follows: *volume scan*: scan time = 15 s (breath–hold), $T_R/T_E = 4.6/2.3$ ms, FOV = 350 mm, slice thickness = 10 mm, image matrix = 256×205; *motility scan*: three parallel image slices each consisting of 120 dynamics, interleaved acquisition, scan time = 155 s (free breathing), $T_R/T_E = 4.0/1.8$ ms, image matrix = 256×205, SENSE factor = 1.6. Six rectangular surface coils (height = 20 cm, width = 10 cm) were fixed around the abdomen and connected to six separate receive channels for signal detection.

On the 0.5 T MRI system, fast spoiled gradient echo (FSPGR) sequences were used for imaging. The imaging parameters for this system were as follows: *volume scan*: scan time = 44 s (two breath–holds), $T_R/T_E = 68/5.6$ ms, FOV = 350 mm, slice thickness = 10 mm, image matrix = 256×160; *motility scan*: 50 – 70 oblique coronal dynamics, scan time = 109 – 148 s (free breathing), $T_R/T_E = 12.5/5.6$ ms, image matrix = 256×160. An abdominal send/receive coil was wrapped around the abdomen for signal detection.

Image Analysis

The total gastric area was outlined in each image slice of a *volume scan*. Based on this segmentation the stomach volume, meal volume and intragastric meal distribution were determined over time. The stomach volume was calculated by summing the outlined pixels in each image slice and integrating the sum over all slices. In each image slice the meal contents could be identified by the distinct positive signal intensity compared to intragastric air (Figure 3.1). Summing the pixels reflecting meal contents and integrating the sum over all slices resulted in the meal volume. Intragastric air volume was determined by subtracting meal volume from stomach volume. A three–dimensional representation of the stomach based on the outlined contours was used to separate stomach volume into proximal and distal gastric volume (Figure 3.2).

The stomach volume; total, proximal, and distal meal volume; and intragastric air volume (expressed in ml) were plotted over time. The complete study period of 90 min was divided in two phases: the ingestion phase (time points $t = -15, -4$, and 0 min) and the emptying phase (time points $t = 0 - 90$ min). Furthermore, the total emptying phase was subdivided into early (0 – 30 min) and late (30 – 90 min) post–prandial periods.

Gastric relaxation was defined as the volume difference between stomach volume directly after and just before meal intake. Gastric relaxation was calculated after solid ($t = -4$ min) and liquid ($t = 0$ min) meal ingestion. Gastric emptying was defined as the decrease in post–prandial meal volume over time. Gastric emptying curves, which were defined as the normalized remaining meal volume in the stomach and expressed as a percentage, were plotted over time. Intragastric meal distribution was defined as the ratio of the distal to the proximal meal volume. Gastric peristalsis was analyzed along a user–defined distal stomach axis. Figure 3.3 shows in detail the analysis of the motility image data.

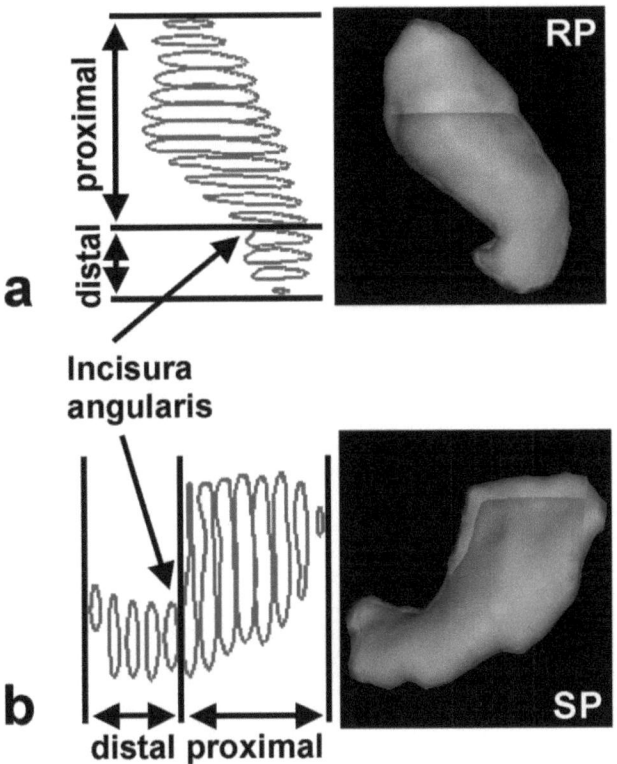

*Figure 3.2: Three–dimensional stomach contours (left) and corresponding volume rendering of stomach and meal volumes (right) 5 min after meal intake for RP (**a**) and SP (**b**). Stomach volume was separated into proximal and distal gastric volumes at the incisura angularis on the lesser curve of the stomach (left).*

Equally spaced profile lines were defined perpendicular to the axis (20 profiles in this example). The signal intensities along each profile of all dynamics were stacked and represented as the "motility plot". The frequency and velocity of gastric contraction waves were determined from these motility plots. Peristaltic frequency and velocity averaged over all time points was calculated. Gastric activity was defined as the mean peristaltic frequency.

Image analysis was performed using an in–house–written software package implemented in IDL 5.5 (Research Systems Inc., Boulder, CO, USA).

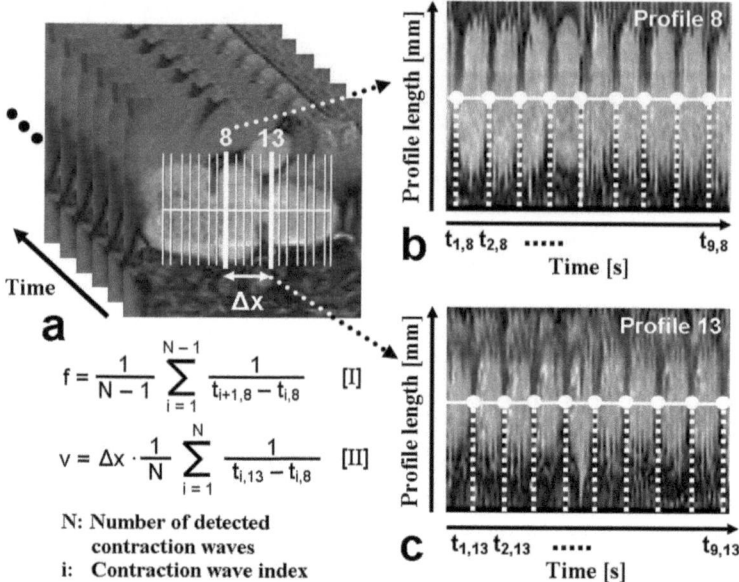

Figure 3.3: *Image series of a representative motility scan displaying distal stomach axis and profiles (**a**). The 8th and 13th profiles separated by the distance Δx are highlighted. The image intensities along the profiles for each dynamic are stacked to produce the motility plots of the 8th (**b**) and 13th (**c**) profiles. The white points indicate the detection of the contraction waves. From the corresponding time points the mean peristaltic frequency f [I] and velocity v [II] are calculated.*

Statistical Analysis

Data were analyzed using SPSS® for Windows Release 11 software (SPSS Inc., Chicago, IL, USA). Logarithmic transformation was applied to stomach, meal and intragastric air volume to normalize data distribution. For four subjects in both RP and SP, volume measurements at $t = 90$ min could not be performed due to technical problems. The last observation carried forward (LOCF) approach was used to fill the missing data points. For the ingestion phase, gastric relaxation and stomach, meal and intragastric air volume were analyzed using a paired Student's t–test. For the emptying phase, a two–factor (body position and time) repeated–measures analysis of variance (ANOVA) with Greenhouse–Geisser correction was used to evaluate the effect of body position and time on total, proximal, and distal meal volume, as well as stomach and intragastric air volume. Analyzing the interaction between body position and time allowed to assess differences in the characteristics of volume curves over time between RP and SP. In addition, meal volume difference at time $t = 90$ min was analyzed using a paired Student's t–test. Gastric activity and mean peristaltic velocity were also analyzed using a paired Student's t–test. Data are expressed and presented as the median (interquartile range). A P–value < 0.05 was considered statistically significant.

Results

The study and both body positions were well tolerated by all subjects. Image acquisition and analysis was successfully performed in all subjects. The image quality attained with both the whole–body and the open–configuration MRI system allowed semiautomated detection and computation of stomach, meal, and intragastric air volume, as well as peristaltic frequency and velocity.

For the ingestion phase, stomach and intragastric air volume was similar for RP and SP for the fasted condition, after solid meal intake, and after liquid meal intake (Table 3.1). Gastric relaxation showed no difference between RP and SP after both solid and liquid meal ingestion (Table 3.1). For the emptying phase, stomach volume, meal volume, and intragastric air volume (expressed in ml) over time is presented in Figure 3.4. Gastric emptying curves are presented in Figure 3.5. Proximal and distal meal volume (expressed in ml) over time is presented in Figure 3.6.

Table 3.1: *Gastric volumes and relaxation during ingestion phase for RP and SP*

	RP			SP		
	Fasted	After solid	After liquid	Fasted	After solid	After liquid
SV (ml)	92 (76–142)	312 (289–353)	458 (446–488)	82 (75–125)	310 (271–344)	461 (435–489)
AV (ml)	69 (53–86)	97 (52–149)	110 (81–131)	67 (51–107)	100 (81–141)	98 (74–147)
GR (ml)		225 (184–244)	163 (118–176)		230 (184–254)	153 (126–182)

RP = right decubitus body position, SP = seated body position,
SV = stomach volume, AV = intragastric air volume, GR = gastric relaxation.
Data are expressed as median (interquartile range).

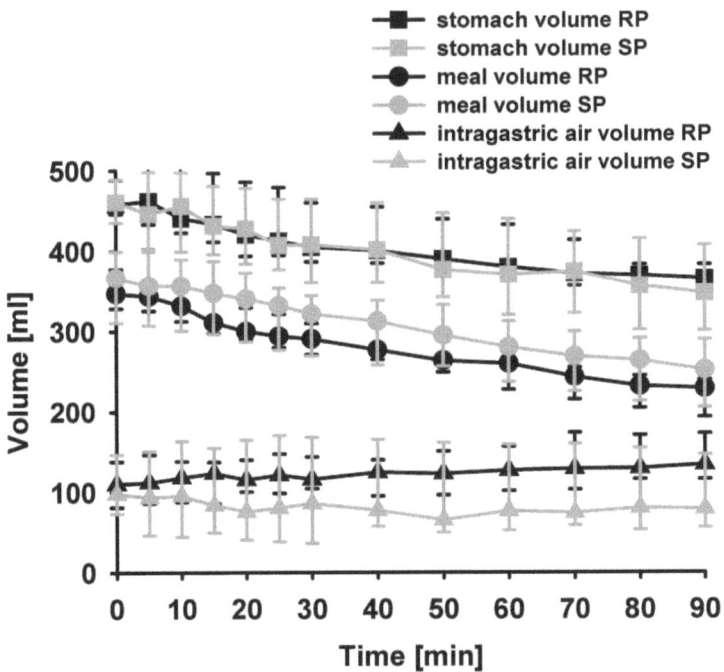

Figure 3.4: Stomach volume (■), meal volume (•), and intragastric air volume (▲) over post–prandial study period for RP (black) and SP (gray). Data are expressed as the median (interquartile range). The large interquartile ranges reflect the high interindividual variability commonly observed for gastric emptying (and not measurement errors).

Gastric Volumes

Maximal stomach volume in RP was reached at $t = 5$ min and thereafter continuously decreased over time (Figure 3.4). In SP stomach volume reached a maximum after complete meal ingestion at $t = 0$ min. However, not until time $t = 10$ min was a continuous decrease observed (Figure 3.4). Statistical analysis showed a significant effect of time ($P < 0.001$), a nonsignificant effect of body position, and a nonsignificant interaction between body position and time.

Gastric emptying curves showed different emptying dynamics for RP and SP, especially during the early post–prandial phase (Figure 3.5). Statistical analysis showed a significant effect of time ($P < 0.001$), a nonsignificant effect of body position, and a significant interaction between body position and time ($P < 0.05$) over total post–prandial period. During the early post–prandial period, statistical analysis showed a significant effect of time ($P < 0.001$), a

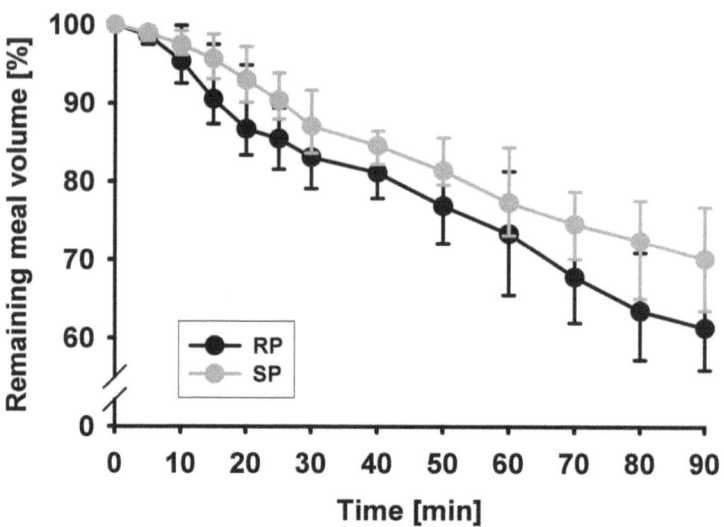

Figure 3.5: Normalized gastric emptying curves (percentage of remaining meal volume in the stomach) for RP (black) and SP (gray). Data are expressed as the median (interquartile range). A divergence of the emptying curves is observed, especially during the first 30 min after meal ingestion.

nonsignificant effect of body position, and a significant interaction between body position and time ($P < 0.05$). During the late post–prandial period, statistical analysis showed a significant effect of time ($P < 0.001$), a trend for an effect of body position ($P = 0.06$), and a significant interaction between body position and time ($P < 0.05$). The remaining meal volume at $t = 90$ min was significantly smaller for RP compared to SP. The (negative) volume difference at this time point was − 43 ml (− 73 ml to − 2 ml, $P < 0.05$).

Initial intragastric air volume at time $t = 0$ min was similar for RP and SP. Subsequently, intragastric air volume curves diverged over time (Figure 3.4). Statistical analysis showed a nonsignificant effect of time, a significant effect of body position ($P < 0.05$) and a trend for an interaction between body position and time ($P = 0.06$) over total post–prandial period. During the early post–prandial period, statistical analysis showed a nonsignificant effect of time, a trend for an effect of body position ($P = 0.07$) and a significant interaction between body position and time ($P < 0.05$). During the late post–prandial period, statistical analysis showed a nonsignificant effect of time, a significant effect of body position ($P < 0.05$) and a nonsignificant interaction between body position and time.

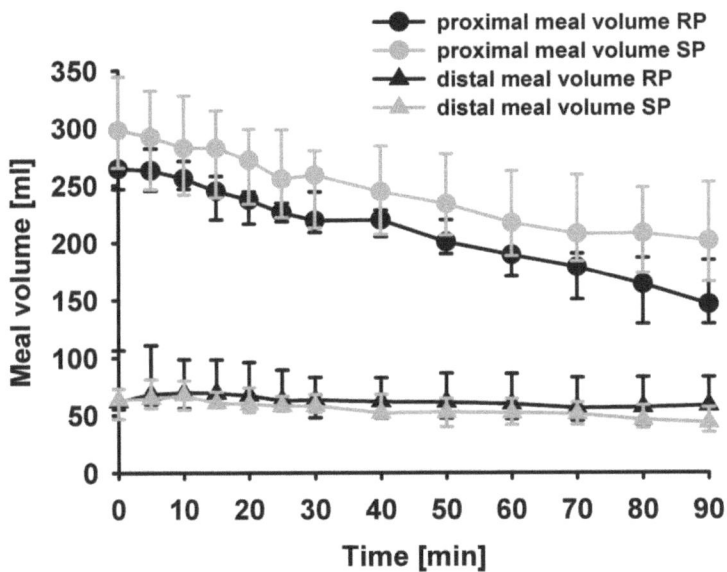

Figure 3.6: Proximal (•) and distal (▲) meal volumes over a post–prandial study period for RP (black) and SP (gray). Data are expressed as the median (interquartile range). Proximal meal volume differed between the two body positions and showed different dynamics for the total emptying phase. Distal meal volume was similar for RP and SP and remained approximately constant over time.

Intragastric Meal Distribution

Proximal meal volume curves reflected the characteristics of total meal volume curves for both positions (Figure 3.6). For proximal meal volume, statistical analysis showed a significant effect of time ($P < 0.001$), a significant effect of body position ($P < 0.05$), and a significant interaction between body position and time ($P < 0.05$). For distal meal volume, statistical analysis showed a significant effect of time ($P < 0.001$), a nonsignificant effect of body position, and a nonsignificant interaction between body position and time. Statistical analysis of intragastric distribution showed a trend for an effect of time ($P = 0.06$), a significant effect of body position ($P < 0.05$), and a significant interaction between body position and time ($P < 0.05$).

Gastric activity did not differ between RP and SP (RP: 3.1 min^{-1} (3.0 – 3.2 min^{-1}) vs. SP: 3.1 min^{-1} (2.9 – 3.3 min^{-1}), n.s.). There was a small difference in peristaltic velocity for the two body positions (RP: 2.6 mm/s (2.3 – 3.3 mm/s) vs. SP: 2.4 mm/s (2.2 – 3.0 mm/s), $P < 0.05$). For both positions, no correlation was found between peristaltic velocity and gastric emptying.

Discussion

This study demonstrates the effect of body position on gastric motor function as assessed by MRI. This is the first study to simultaneously analyze the posture dependency of eight relevant gastric parameters. Stomach, meal, and intragastric air volume; gastric relaxation and emptying; intragastric meal distribution; and peristaltic frequency and velocity were successfully assessed in RP and SP using different MRI systems. The results indicate that gastric MRI in RP is a valuable and noninvasive imaging technique for assessing human gastric function.

Gastric volume responses during the ingestion phase were not altered by RP, and stomach volume remained similar throughout the complete study period for the two body positions. An effect of body position was found for post–prandial intragastric air volume. This effect was explained by significant different dynamics in the meal emptying between RP and SP. These different emptying dynamics produced a trend for an effect of body position on late post–prandial meal volume, and finally resulted in a slight meal volume difference of -43 ml at time $t = 90$ min.

The effect of posture on gastric function was investigated in several studies using γ–scintigraphy,[34-38] intraluminal manometry,[35] and MRI.[95] Because of the different techniques and various test meals used, the influence of body position on the emptying of a solid/liquid meal remains controversial.[36,37] More importantly, none of these studies simultaneously analyzed the effect of posture on the dynamics of gastric emptying and related peristaltic activity. However, previous studies clearly demonstrated the strong impact of intragastric fat distribution (fat layering) on gastric emptying.[34,95] Therefore, in the present study a low–fat solid/liquid meal and the RP body position was used to minimize this effect.

Gastric emptying was significantly different for RP and SP. However, the difference in emptying dynamics and meal volume after 90 min was subtle and small compared to the changes observed for meal emptying in patients diagnosed with gastroparesis.[96] Since gastric activity (mean peristaltic frequency) was similar for both positions, and the slightly faster peristaltic velocity for RP did not correlate with the patterns of emptying, we conclude that gastric peristalsis did not accelerate meal emptying in RP. This is in agreement with the results of previous studies that used MRI[97] and duplex sonography[98] combined with intraluminal manometry. Those studies showed that a gastroduodenal pressure difference, rather than gastric peristalsis, may be the major driving force for emptying. Gastroduodenal pressure difference is determined by the intragastric pressure (IGP), which is influenced by the intraabdominal pressure

(IAP) and gastric wall tension. In this noninvasive study, no *in vivo* pressure or strain measurements were performed to detect these parameters. However, differences in IGP, IAP, and gastric tone between the two body positions can be assumed indirectly from gastric volumes. In respiratory medicine, as well as in studies using the gastric barostat (a "gold standard" for assessing gastric tone), IAP is approximated from fasted IGP.[99,100] Based on the thermodynamic interrelationship between air pressure and air volume ($p \cdot V = const.$), a comparison of fasted intragastric air volume allows the indirect determination of possible differences in IAP between RP and SP. Since fasted intragastric air and stomach volume did not differ between the positions, IAP was assumed to be similar for RP and SP during the complete study period. This assumption implies that initial post–prandial IGP must have been higher in RP, since the emptying curve diverged over the emptying period from patterns observed in SP (Figure 3.5), with no alteration in the dynamics of stomach volume. Comparable initial meal and stomach volumes indicated that increased IGP was generated by higher intragastric air pressure. MRI volume data for RP, and a corresponding three–dimensional visualization (Figure 3.2a) showed that gastric contents were located above the lower esophageal sphincter (LES), which prevented belching out of the intragastric air. Thus, swallowed air during liquid ingestion was trapped inside the proximal stomach in RP and may have caused an increase in IGP. This idea is indirectly supported by the higher interindividual variability in intragastric air volume for SP, a body position in which it is generally easier to expel air from the stomach (Figure 3.4). Obviously, a higher IGP in RP could only be maintained by increased gastric wall tension (increased gastric tone), which can be induced by an increased vagal stimulation in the lying position,[101–103] especially in RP.[104,105]

Another factor to consider as a major control mechanism for gastric emptying, apart from alterations in IGP and gastric tone (and gastric peristalsis), is the resistance to gastric outflow provided by the pylorus.[98,106,107] In animals, pyloric function is also modulated by vagal activity. Vagal stimulation reduces pyloric resistance and increases transpyloric stroke volume,[108,109] whereas vagal blockade induces the opposite effect.[110,111] Hence, the effect of vagal stimulation on gastric tone (increase) and pyloric resistance (reduction) indicates that vagal stimulation may be the physiological cause of the accelerated meal emptying in RP.

The current study demonstrates that MRI is a sensitive, noninvasive imaging technique for measuring gastric volume and motor function. Despite large interindividual variations (Figure 3.4), small differences in gastric emptying were detected. Such high sensitivity to subtle differences in the emptying characteristics is currently not attainable with γ–scintigraphy and single photon emission computed tomography (SPECT). Although these techniques are considered the "gold standard" for gastrointestinal imaging, they have many disadvantages compared to MRI. Scintigraphy is restricted to two-dimensional imaging of poor image resolution and requires corrections for background noise and decay. The spatial and temporal resolution of dynamic scintigraphy is very low compared to dynamic MRI (scintigraphy: 5×5 mm^2 at 20 frames/min vs. MRI: 1.4×1.4 mm^2 at 140 frames/min).[112] SPECT measurements require longer acquisitions of gastric volume ranging from 3 to 7 min at lower image resolution of 3×3 mm^2.[113–115] In the current study using MRI, complete volume acquisition was achieved within 15 – 44 seconds at an image resolution of 1.4×1.4 mm^2. Furthermore, neither separation of gastric contents and intragastric air nor simultaneous assessment of gastric volume and peristalsis is feasible using nuclear imaging techniques.

In conclusion, gastric MRI performed in RP is feasible for clinical GI research on gastric motor function. The presented imaging technique has the potential to complement and specify the diagnosis of gastroparesis and motility disorders. The physiological results for IGP, gastric tone, and pyloric resistance indicated that posture-dependent vagal activity induced the different emptying characteristics observed for RP and SP. MRI is a highly sensitive and noninvasive imaging technique that has the potential to become the method of choice for assessing GI function in humans.

Chapter 4

Effect of Gastric Secretion on Gastric Physiology and Emptying in the Fasted and Fed State Assessed by Magnetic Resonance Imaging

Introduction

Gastric secretion is of key importance for the peptic digestion of food and has a role in regulating the rate of gastric emptying and delivery of nutrients to the small intestine. Furthermore, gastric secretion represents an important factor in the pathogenesis of common gastrointestinal diseases including peptic ulcer and gastroesophageal reflux disease (GERD).[23] The focus of previous investigations has been on measuring hydrogen ion concentration (pH) in the gastrointestinal tract; however, (i) pH alone does not reflect the absolute "amount" of acid produced by parietal cells in the stomach or present in the refluxate, (ii) gastric secretion contains other potentially harmful chemicals (e.g. pepsin, trypsin and bile acid), and (iii) volume reflux together with esophageal distension is a common cause of ongoing symptoms in GERD patients that fail to respond to acid suppression with proton pump inhibitors (PPIs).[116,117] The important contribution of gastric secretion to post–prandial intragastric volume is demonstrated by reports that approximately 800 ml of gastric juice is produced in response to a 400 ml solid meal, with the secretory profiles of acid and pepsin showing a similar dynamic.[118,119]

Current measurement techniques for gastric secretion involve gastric and/or intestinal aspiration or *in vivo* titration of gastric contents before and after a meal (sometimes with pentagastrin stimulation).[75,76,118,120] These methods are invasive, unphysiological and may themselves alter gastric secretion.[118,121] Moreover, both are applicable only with experimental, mainly liquid test meals and allow only a global determination of gastric secretion although it is known that mixing of gastric contents is poor and local distribution of gastric secretion is often inhomogeneous (e.g. layering).[23,77] In addition, these techniques provide no information about the effects of gastric secretion on gastric physiology. Indeed, no current method provides this information. Scintigraphy assesses only the fractional emptying of ingested radio–labeled test meals, barostat and single photon emission computed tomography (SPECT) measure total gastric volume change (wall relaxation and contraction (tone)) and gastric ultrasound quantifies gastric content volume; however, none of these techniques can assess the dilution and mixing of the test meal with gastric secretion or the interplay between meal volume, gastric secretion and gastric volume response.

The ideal measurement technique would be safe, noninvasive and provide an accurate assessment of the volume and distribution of gastric secretion in the fasted and fed states. It would also allow the effects of gastric secretion on gastric motor function, intragastric distribution of the meal and gastric emptying to be assessed. Magnetic resonance imaging (MRI) has many of these properties. Marciani and colleagues presented MRI measurements of gastric secretion based on *in vitro* calibration of inverse T_2 relaxation time against the polysaccharide concentration of a specific viscous test meal.[79,80] Recently, a fast T_1 mapping technique for the exact quantification of intragastric dilution and distribution of orally applied gadolinium (Gd) based paramagnetic contrast agents (CAs) was developed that can be applied to a variety of test meals.[81] Extending this work, quantification of intragastric secretion by calibrating T_1 against CA concentration was validated in healthy volunteers.[122]

The current study applied this noninvasive technique to investigate the effect of gastric secretion on gastric physiology in the fasted and fed states. Pentagastrin stimulated secretion was compared to placebo in a randomized, double–blind controlled cross–over experimental study design. Direct MRI measurements of gastric secretion were related to the dynamic effects of pentagastrin stimulated secretion on gastric content volume, total gastric volume (i.e. gastric tone) and gastric emptying.

Materials and Methods

The study was carried out according to Good Clinical Practice and the Declaration of Helsinki. Written informed consent was obtained from all participants and the protocol was approved by the local Ethics Committee and the Swiss national agency for therapeutic products. All measurements were completed without complications or adverse events.

Subjects and Study Design

Twelve healthy subjects (five women and seven men; mean age = 28 years (range = 22 – 34 years); body mass index (BMI) = 23.2 ± 2.0 kg/m^2) participated in this double–blinded, randomized, placebo–controlled, cross–over study. None of the volunteers had gastrointestinal symptoms, previous abdominal surgery (except appendectomy), or was taking any medications apart from oral contraceptives. All subjects had a similar dietary history in the last two weeks. Each subject was investigated after an eight–hour fasting period on four occasions separated by 2 – 14 days according to the randomized double–blind study design.

In study sequence A, the effect of pentagastrin on gastric juice production in fasted state was assessed. In study sequence B, the effect of pentagastrin on gastric juice production after ingestion of a caloric 500 ml liquid viscous test meal was assessed. The study order was randomized both between and within the sequences. Each volunteer ingested the test meal in 3 – 5 min. Continuous intravenous pentagastrin (0.6 µg/kg/h; Cambridge Laboratories, UK) or placebo (NaCl 0.9 % Baxter®; Baxter, Volketswil, Switzerland) infusion started with the meal and continued for 60 min under electrocardiogram (ECG) and blood pressure monitoring. MRI measurements continued for a further 30 min.

Test Meal

Glucose solution (500ml, 100 g/l (200 kcal); Fresenius Kabi AG, Switzerland) was homogeneously mixed with 5 g locust bean gum (LBG) powder (Rapunzel Naturkost AG, Germany) at 37°C. Obtained test meal was labeled with 1.2 ml of the paramagnetic contrast agent Gadolinium–DOTA (Gd–DOTA; Dotarem®, Laboratoire Guerbet, France) resulting in a contrast agent concentration C_{Gd} of 1200 µM. Secretion volume was quantified by measuring T_1 values of the diluted test meal and using the interrelationship between T_1 and C_{Gd} (Figure 4.1) as described in detail in a previous study.[122] Gd–DOTA complexes remained stable in the acid environment over the complete study duration.[29]

Magnetic Resonance Imaging

MRI measurements were performed in supine body position on a 1.5 T whole–body MRI system (1.5 T Achieva, Philips Medical Systems, Best, The Netherlands) using an abdominal, four–channel phased–array receive coil for signal detection.

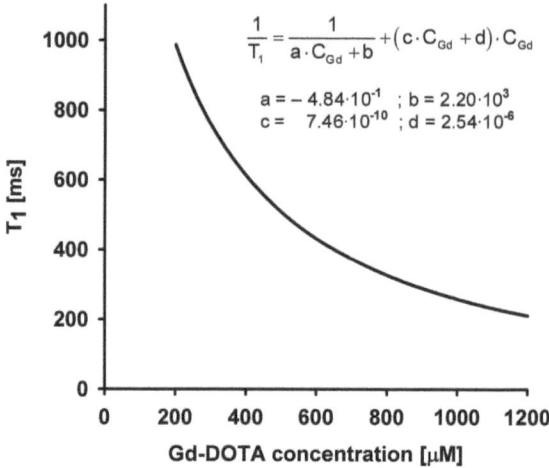

Figure 4.1: Calibration curve describing the interrelationship between longitudinal relaxation time T_1 and contrast agent concentration C_{Gd} of the test meal that was continuously diluted with gastric secretion. Corresponding equation for T_1 and C_{Gd} (expressed in ms and µM, respectively) with empirically derived parameters a, b, c, and d is shown. This algorithm allows the dilution of labeled gastric contents to be assessed from T_1 maps.

Study Sequence A (fasted state)

The complete gastric region was imaged (volume scan) at baseline to assess initial gastric content volume ($GCV_{residual}$) and initial total gastric volume ($TGV_{residual}$). At time $t = 0$ min (t_0) the pentagastrin/placebo intravenous infusion was started and continued for 60 min. Volume scans were performed every 5 min for the entire 90 min study period.

Study Sequence B (fed state)

A volume scan was performed at baseline. The pentagastrin/placebo intravenous infusion was started at $t = -10$ min (t_{-10}) and subjects ingested the liquid nutrient test meal in supine position. Immediately after meal ingestion at $t = 0$ min (t_0) and thereafter in 5 min intervals for the entire 90 min study period, volume scans followed by combined T_1 mapping (to generate T_1 maps) and B_1 mapping scans (to correct for radiofrequency inhomogeneities) were performed. Sequence parameters have been published in the validation of the presented technique.[122]

Image Analysis

Volume Data

Total gastric volume (TGV) and gastric content volume (GCV) were assessed from MR images of volume scans by a semi–automatic analysis as described previously.[123] The difference between TGV and GCV resulted in intragastric air volume. Repeated measurements provided a dynamic assessment of volume change and gastric emptying over time.

T_1 Maps

T_1 maps were computed for every combined T_1 and B_1 mapping scan using an algorithm validated in a previous study.[122] Since T_1 values of undiluted meal components are normally distributed around the reference T_1 value of 211 ms,[122] T_1 distribution of diluted meal was obtained by subtracting the normal distribution of the undiluted meal from the overall T_1 distribution of gastric content. Layering of gastric juice was defined as diluted meal observed at the interface between intragastric air and gastric content on T_1 maps with a minimum thickness of at least two pixel lines (to prevent errors arising from susceptibility artifacts between air and meal). Layer volume was calculated by summing all pixel areas reflecting gastric juice layer and multiplying the sum by the slice thickness. Image analysis was performed using an in–house–written software package implemented in IDL 6.2 (Research Systems Inc., Boulder, CO, USA).

Data Analysis of Gastric Volume

In study sequence A, the study period was sub–divided into two phases, the infusion phase ($t_0 - t_{60}$) and the post–infusion phase ($t_{60} - t_{90}$). The area under the gastric content (i.e. secretion) volume curve (AUC [l·min]) was calculated for both time periods using the trapezoid method. The half emptying time (T_{50}) of the gastric content volume after stopping the pentagastrin infusion was calculated using a power exponential model.

Study sequence B was sub–divided into two phases, the ingestion phase ($t_{-10} - t_0$) and the gastric emptying phase ($t_0 - t_{80}$). The gastric emptying phase was further sub–divided into the infusion phase ($t_0 - t_{50}$) and the post–infusion phase ($t_{50} - t_{80}$). Gastric relaxation ($TGV_{filling}$) was defined as the volume difference between the stomach volume post–ingestion at $t = 0$ min and the volume at baseline ($TGV_0 - TGV_{residual}$), initial meal emptying as the difference of ($GCV_{residual}$ + nominal meal volume (500 ml)) and GCV_0. Stomach and meal volume was plotted over time to generate volume curves describing the dynamic gastric volume response (relaxation and contraction) and gastric emptying over time. The area under the volume curve (AUC [l·min]) for TGV, GCV, layer volume and intragastric air volume was calculated using the trapezoid method.

To analyze the characteristics of the volume curves, the data was fitted to a three–parameter gastric emptying model. The model formula is given as:

$$V(t) = V_0 \cdot \left(1 + \frac{\kappa \cdot t}{t_{empt}}\right) \cdot exp\left(-\frac{t}{t_{empt}}\right)$$

where V_0 [ml] is an estimate of the post–prandial volume at t_0 and t_{empt} is the emptying time constant in [min]. The dimensionless positive parameter κ models dynamic volume change. The special case of $\kappa = 0$ describes exponential emptying, whereas $\kappa > 1$ describes a volume rise after $t = 0$ min. This model was introduced recently[78, 124] to assess the limitations of the power exponential gastric emptying model for normalized volume data.[125] The gastric emptying model presented can describe the increase in volume often observed after meal ingestion in gastric MRI studies, whereas the power exponential model is limited to monotonically decreasing gastric emptying curves (Figure 4.2).

Statistical Analysis

For study sequence A, volume data over time was analyzed using paired repeated–measures analysis of variance (ANOVA). If a significant interaction of treatment and time was observed, values at single time points were compared by one–way ANOVA.

For study sequence B, volume data was analyzed applying the novel gastric emptying model. A single statistical fit was performed using the library *nlme* of the data analysis package R[126, 127] to stabilize parameter estimates (V_0, κ, t_{empt}). T_{50} for stomach and meal volume was determined from κ and t_{empt} by Newton approximation. Maximal gastric emptying rate (GER_{max}) was calculated by derivation of the gastric emptying model curves. Parameters obtained from the model were compared using a mixed effect model ANOVA with "subject" as random variable and "treatment" and "meal/stomach" as fixed variables.[126] For both study sequences single parameter comparisons were analyzed using a paired Wilcoxon test. Data was considered to be statistically significant at $P < 0.05$ and was presented as mean ± SD. All statistical calculations were carried out using the data analysis package R and SPSS 10.0.7 for Windows.[127]

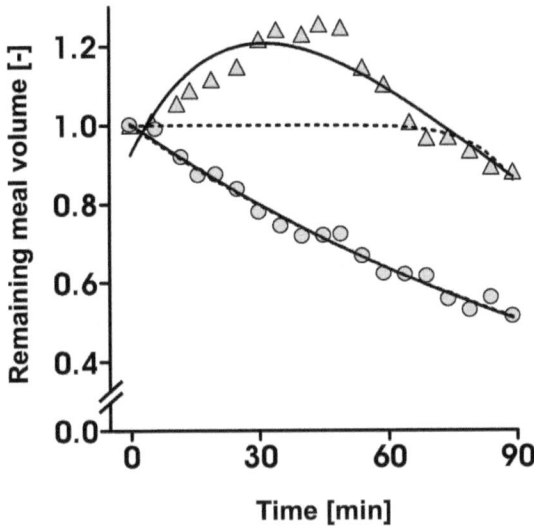

Figure 4.2: Normalized gastric content volume (fraction of remaining meal volume in the stomach) of a representative healthy volunteer during intravenous infusion of placebo (circles) and pentagastrin (triangles). Volume data was fitted by a standard power exponential emptying model (dashed line) and by the novel gastric emptying model (solid line). The standard power exponential fit does not describe the initial volume increase induced by pentagastrin and thus cannot be used to analyze the interaction between secretion and emptying that define gastric content volume at any given time point.

Results

Study Sequence A (fasted state)

There was no difference in $GCV_{residual}$ between placebo and pentagastrin (40 ± 29 ml vs. 35 ± 19 ml, n.s.). Placebo had no effect on gastric content volume over 90 min (slope: 0.0018 ml/min, n.s.). Pentagastrin induced a continuous increase in gastric content volume (significant after 15 min) resulting in a maximum volume of 119 ± 35 ml compared to 32 ± 20 ml for placebo ($P < 0.001$) at 60 min (Figure 4.3a, b). After stopping the pentagastrin infusion gastric content volume emptied with a mean half emptying time T_{50} of 34 ± 14 min. The different dynamic of GCV over the study period was confirmed by AUC_{0-60} and AUC_{60-90} for placebo and pentagastrin (2.1 ± 1.2 vs. 4.6 ± 1.3 and 0.8 ± 0.5 vs. 2.6 ± 1.1 l·min, $P < 0.001$ for both $AUCs$).

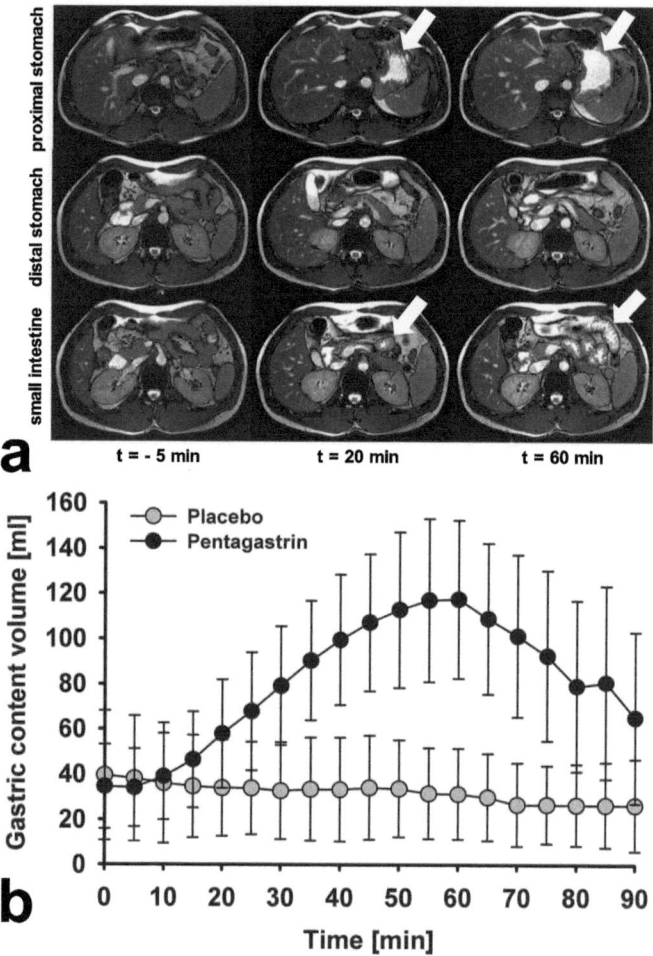

Figure 4.3: **(a)** Transverse MR image slices of a healthy volunteer of the proximal (top row) and distal stomach (middle row) and the small intestine (bottom row) at three different time points ($t = -5$ min, left column; $t = 20$ min, middle column; $t = 60$ min, right column). Gastric content volume increased from 20 ml at $t = -5$ min to 140 ml at $t = 60$ min during pentagastrin infusion (indicated by arrows in top row). Simultaneously, gastric content was emptied in the small intestine (indicated by arrows in bottom row). **(b)** Mean ± SD of gastric content volume for placebo (gray) and pentagastrin (black) of 12 healthy volunteers.

Study Sequence B (fed state)

Pre–prandial ($TGV_{residual}$, $GCV_{residual}$) and initial post–prandial total gastric volume and gastric content volume (TGV_0, GCV_0) was not different for placebo and pentagastrin (Table 4.1). Also, there was no difference in gastric relaxation and initial meal emptying between placebo and pentagastrin (Table 4.1).

Table 4.1: *Gastric volume responses during i.v. infusion of placebo and pentagastrin*

	Placebo		Pentagastrin		Unit
	GCV	TGV	GCV	TGV	
$V_{residual}$	25±10	107±25	30±24	104±41	[ml]
V_0	478±35	593±61	494±34	602±58	[ml]
$V_{filling}$	–	485±46	–	499±55	[ml]
$V_{emptying}$	48±34	–	35±37	–	[ml]
κ	0.6±0.3	0.7±0.3	1.6±0.3††	1.6±0.3††	–
T_{50}	53±19	70±24	96±53††	110±47††	[min]
GER_{max}	5.9±2.1	5.4±1.8	4.9±1.5	5.1±1.5	[%/min]
AUC_{inf}	16±2	21±3	22±4††	27±4††	[l·min]
$V_{max,L}$	25±24	–	99±56†	–	[ml]
AUC_L	0.7±0.8	–	4.7±3.1†	–	[l·min]

GCV	=	gastric content volume
TGV	=	total gastric volume
$V_{residual}$	=	pre-prandial volume
V_0	=	initial post–prandial volume
$V_{filling}$	=	gastric relaxation
$V_{emptying}$	=	initial meal emptying
κ	=	model parameter describing the increase in gastric volume
T_{50}	=	model parameter describing the half emptying time
GER_{max}	=	maximum gastric emptying rate
AUC_{inf}	=	area under the volume curve during infusion
$V_{max,L}$	=	maximum layer volume of gastric secretion
AUC_L	=	area under the volume curve of gastric secretion over total study period

† $P < 0.01$; †† $P < 0.0001$.
Data are expressed as mean ± SD.

TGV was always larger than GCV; however, the dynamic change in TGV and GCV over the period of the study was very similar (Figure 4.4), as indicated by similar values for the dimensionless model parameter κ for both measurements in both pentagastrin and placebo study conditions (Table 4.1). Consistent with this finding, intragastric air volume was also not different between the conditions (data not shown).

TGV and GCV showed a distinct increase in volume during early phase of pentagastrin infusion which was less pronounced during placebo infusion, before beginning to fall at an approximately linear rate in both study conditions (Figure 4.4). The early rise in GCV was associated with the development of a layer

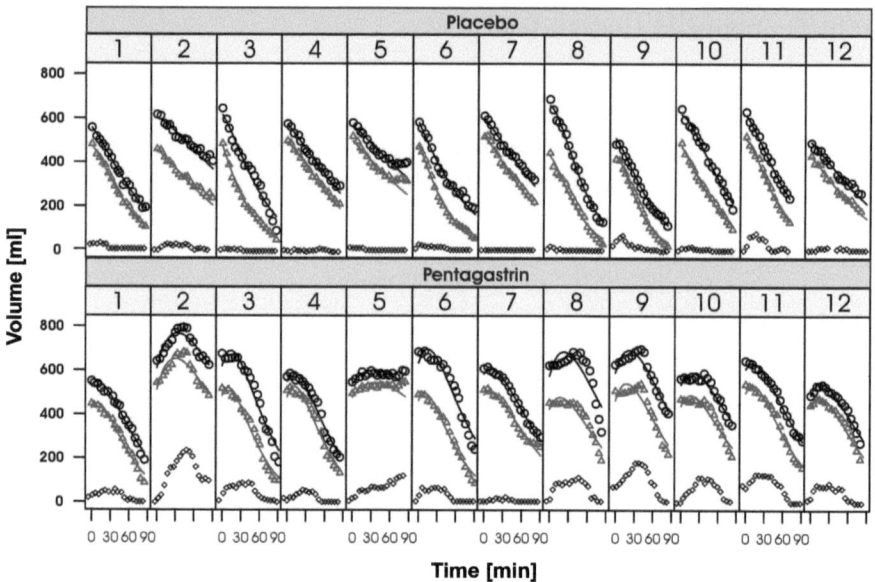

Figure 4.4: Individual volume data for total gastric volume (TGV, circles), gastric content volume (GCV, triangles) and layer volume of gastric secretion (diamonds) in the placebo (top) and pentagastrin (bottom) experimental conditions. For TGV and GCV curve fits derived from the gastric emptying model are plotted. Note the similar dynamics of the pentagastrin induced layering process, TGV and GCV.

of gastric secretion on top of the test meal in the proximal stomach that was clearly observed on T_1 maps (Figure 4.5). The effect of stimulated secretion on the volume curves is reflected by the significantly higher value of the parameter κ and AUC for GCV during pentagastrin compared to placebo infusions (Table 4.1). Moreover, there was a highly significant linear relationship between the dynamics of GCV change over the course of the study (described by model parameter κ) and direct measurements of maximal gastric secretion layer volume ($V_{max,L}$) and AUC_L as shown in Figure 4.6. In addition to greater secretion layer volume, the mean contrast agent (CA) concentration in the layer of all volunteers was more dilute in pentagastrin than placebo conditions (432 ± 59 μM vs. 506 ± 90 μM, $P < 0.001$), which corresponds to a CA dilution ratio of test meal to gastric secretion of 1:2.8 for pentagastrin and 1:2.4 for placebo. In the remainder of gastric contents, mixing of the test meal with gastric secretion was very incomplete. There was a dilution gradient from proximal to distal with a mean CA concentration in the distal stomach of 652 ± 333 μM (dilution ratio of 1:1.8) for pentagastrin and 860 ± 301 μM (dilution ratio of 1:1.4) for placebo ($P < 0.001$). The secretory response was highly variable in both conditions;

however, subjects that had a pronounced secretory response to a meal also had a greater response to pentagastrin and a weak linear association in AUC_L was observed between pentagastrin and placebo ($R^2 = 0.39$, $P = 0.03$).

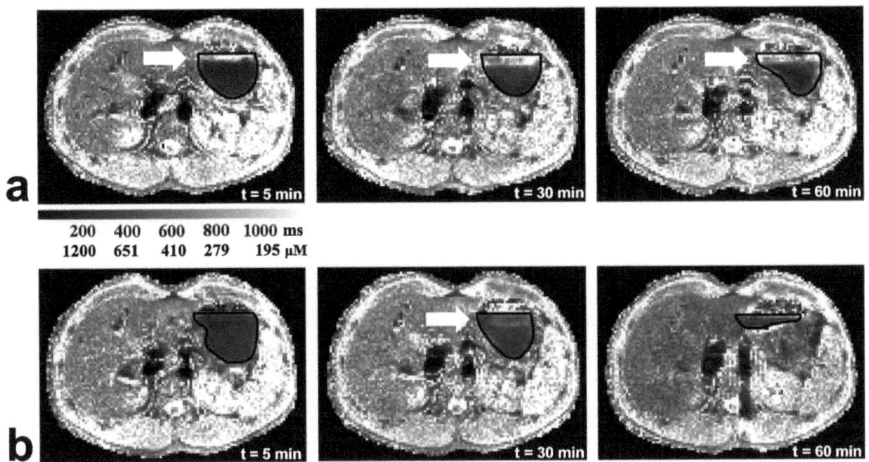

Figure 4.5: B_1 corrected T_1 maps (in ms) encoding intragastric Gd–DOTA concentration (in µM) of a representative proximal image slice of a healthy volunteer with outlined stomach wall. T_1 maps are presented at three different time points (t = 5 min, 30 min, 60 min) for pentagastrin (a) and placebo (b). Layering of gastric secretion is indicated by arrows.

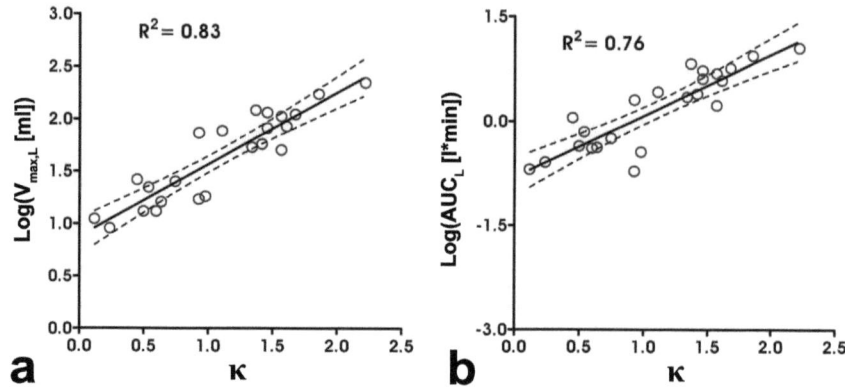

Figure 4.6: (**a**) Linear regression analysis of log–transformed maximal layer volume of gastric secretion ($V_{max,L}$) and model parameter κ, and (**b**) log–transformed area under the curve of the layer volume of gastric secretion (AUC_L) and model parameter κ. The close correlations indicate that the κ coefficient accurately reflects the effect of secretion on gastric content volume over the course of the study.

Comparing pentagastrin and placebo an important discrepancy between gastric emptying of the liquid nutrient test meal as assessed by half emptying time T_{50} and maximal gastric emptying rate (GER_{max}) was evident. The early volume increase during pentagastrin infusion was associated with a prolonged gastric half emptying time T_{50} compared to the placebo condition; in contrast, GER_{max} was not significantly different between the two conditions (Table 4.1).

Discussion

The fast T_1 mapping technique provided direct, noninvasive quantification of gastric secretory products and their distribution (i.e. layering) in the human stomach. Combined with repeated MRI measurement of gastric content volume (GCV) and total gastric volume (TGV), the effects of gastric secretion on gastric function and emptying were described and parameterized by a gastric emptying model. This novel analysis showed that gastric half emptying time (T_{50}) was increased during pentagastrin infusion compared to placebo due to increased secretion (and therefore GCV) with the formation of an intragastric layer of secretory products and not by significant reduction of the gastric emptying rate.

This study provided a vivid description and quantitative assessment of the effects of gastric secretion on the post–prandial gastric response (relaxation and contraction) and gastric emptying. In the fasted state, the volume of residual gastric secretion did not change during the placebo infusion. As expected, pentagastrin greatly increased gastric secretion volume (i.e. GCV) with significant amounts of secretory products collecting in the stomach and also entering the small intestine (Figure 4.3). This secretion volume continued to increase until the pentagastrin infusion was completed (maximal GCV of 95 ml to 171 ml), after which gastric contents emptied in a rapid, exponential manner similar to water. In the fed state, gastric secretion and gastric emptying both commenced with meal ingestion, as evidenced by the development of a layer of highly diluted gastric contents above the labeled meal observed on T_1 maps (Figure 4.5) and the presence of gastric contents in the duodenum, respectively. Characteristic gastric volume responses and emptying were observed (Figures 4.2 and 4.4). During pentagastrin infusion, GCV and TGV increased for a few minutes after meal ingestion before starting to decrease in an approximately linear manner typical of nutrient fluids. This rise in intragastric volume was associated with the volume of the secretion layer that formed on top of the labeled test meal (Figure 4.4). The pronounced increase in GCV was not observed during the placebo infusion and the secretion layer volume was reduced but still present, representing the physiological secretory response to the meal.[128] Thus, in most cases the rate of secretion was greater than the rate of emptying during pentagastrin but not placebo infusion. For both conditions, the dynamic relationship can be expressed at any time point by *gastric content volume = (meal volume + secretion volume) - emptied volume*. The gastric emptying model used in this study parameterized the early dynamic change in gastric volume (including a rise in GCV, if present) by the κ coefficient, and the subsequent decrease in GCV by t_{empt}.[78] This analysis has important implications for the assessment of gastric emptying in terms of half emptying time (T_{50}) because the secretory response introduces a delay before GCV begins to fall below the volume measured immediately after ingestion (t_0). Thus, the increase in GCV during pentagastrin infusion resulted in significantly longer T_{50} even though the maximal gastric emptying rate (GER_{max}) remained unchanged. MRI differentiated these two distinct factors *in vivo*.

The ability to measure changes in TGV and GCV (including meal and secretion volume) simultaneously by noninvasive MRI techniques provides a more physiological and comprehensive assessment of the effects of pentagastrin and gastric secretion on the stomach than invasive methods that focus on individual aspects of function or emptying. For example, barostat experiments suggest that pentagastrin induces gastric relaxation and that this effect is reduced by H2 receptor antagonists or proton pump inhibitors.[129-133] However, this work does not account for the gastric distension produced by increased secretion volume during pentagastrin infusion. Consistent with studies that measured both intragastric volume and gastric relaxation,[133] the current findings demonstrate that changes in TGV during filling and in the post–prandial period during pentagastrin infusion reflect changes in GCV only, indicating that gastric relaxation is driven by intragastric volume and not by a direct effect of gastrin on gastric wall tone. This finding adds to the evidence that barostat studies do not provide a description of gastric volume responses that reflects those that occur under physiological, noninvasive conditions.[78,134] Similarly, the lack of the effect of pentagastrin on gastric emptying rate is at odds with reports that increased gastric acid production and instillation of exogenous acid to the stomach slow gastric emptying in a dose dependent manner.[96,135–137] These findings were based either on aspiration of gastric and duodenal contents without the knowledge that the mixing and distribution of gastric secretion is highly inhomogeneous, or on indirect evidence from manometry (e.g. isolated pyloric pressure waves, MMC phase III frequency in the interdigestive phase).[137–141] The current study demonstrates that stimulation of endogenous gastric acid production by pentagastrin increases gastric half emptying time (T_{50}) due to the GCV increase produced by secretion, but has no significant effect on the rate of gastric emptying per se. Thus, T_{50} depends on the volume of gastric secretion (not appreciated by conventional measurement techniques) as well as the rate of gastric emptying, a fact that clearly limits the usefulness of T_{50} as a single summary endpoint in physiological studies. Further discrepancy between current and previous findings may be due to reduced passage of acid into the small bowel (and enterogastric feedback inhibition) because of layering of secretory products on top of the test meal;[136] however, the presence of a secretion layer does not imply that gastric acid was not present in the remainder of the gastric contents, indeed T_1 mapping revealed dilution of the test meal throughout the stomach.

Several lines of evidence support the contention that the pattern of GCV change observed by MRI in the post–prandial period depends on the dynamic balance between the rate of gastric secretion and the rate of gastric emptying. Firstly, direct measurements of gastric secretion layer volume using the fast T_1 mapping technique correlated with the change in GCV (described by κ in the gastric emptying model) in both placebo and pentagastrin stimulated conditions (Figure 4.6). Secondly, the T_1 values of gastric secretion in this study were identical to that obtained by aspiration in earlier validation studies.[122] Thirdly, although the secretory response was variable, for a given individual the volume of gastric secretion correlated under pentagastrin and placebo conditions. Thus, even without the T_1 mapping technique, direct MRI measurements of GCV provide a surrogate assessment of gastric secretion. It should be emphasized that an increase in GCV did not occur in all patients during pentagastrin infusion and that gastric secretion continued during gastric emptying (layer volume in the placebo phase was maximal at around $t = 50$ min). The effect of secretion on gastric emptying is better described by the gastric emptying model developed specifically for gastric MRI volume measurements (see methods).[78,124] The coefficient κ from this model parameterizes deviation of the gastric emptying curve from a monotonous linear or exponential reduction in volume, with values > 1 indicating a GCV rise in the post–prandial period. This study shows that the coefficient κ reflects the effect of secretion volume on GCV during gastric emptying (Figure 4.6) and, as discussed above, that T_{50} depends on the interaction between the volume of gastric secretion and the rate of gastric emptying after a meal. This dynamic process cannot be described by the "Power Exponential" (PowExp) model of gastric emptying used in many previous studies (Figure 4.2).[125]

This study investigated the effects of gastric secretion on gastric function and emptying using a robust, double–blind, placebo–controlled study design. Pentagastrin stimulation served to highlight the effects of gastric secretion layer volume on post–prandial GCV; effects that were present, but less obvious, also in the placebo phase. The viscous, nutrient liquid meal promoted layering of gastric secretion. For less viscous liquid meals this effect may be reduced due to more efficient mixing of gastric contents; however, the effects on GCV described here have been observed with different test meals.[78] Moreover, the T_1 mapping technique does not depend on layering to provide direct, quantitative measurements of secretion.[122] For solid meals the gastric emptying curve would be further complicated by the presence of a "lag phase" during trituration; however, an initial rise in GCV would still indicate gastric secretion, and layering of secretions is still observed.[55] Homogeneous Gd–DOTA labeling and T_1 mapping of solids may be possible for foods such as pancakes and scrambled eggs (often used in gastric emptying studies); however, it will not be feasible for chunks of meat or vegetables. A shortcoming of this study is that gastric acid was not assessed independent of gastric secretion volume; however, validation studies showed that pH does not have important effects on the T_1 relaxation time.[122] Moreover, although pentagastrin increases both secretion volume and the absolute amount of acid secreted, the pH of gastric secretions varies with secretion volume under (patho)physiological and stimulated conditions.[118,119,142,143]

In conclusion, the fast T_1 mapping technique provided direct, noninvasive measurements of the volume and distribution (i.e. layering) of gastric secretion in the human stomach. These measurements demonstrated that the increase in gastric volumes after a meal often observed in gastric MRI studies is because the volume of gastric secretion was greater than the volume emptying from the stomach in the early post–prandial period. Increased gastric secretion volume drove gastric relaxation; however, there is no evidence that pentagastrin had a direct effect on gastric function. Similarly, increased gastric secretion volume prolonged conventional measurements of gastric emptying in terms of T_{50}; however, the maximal gastric emptying rate (GER_{max}) was not significantly affected. In addition to clarifying the effects of gastrin and gastric secretion on gastric physiology, the ability to visualize the distribution of secretion in the stomach is of interest for studies of gastroesophageal diseases related to gastric secretion (the layer of gastric secretory products above the test meal may be the anatomical correlate of the "gastric acid pocket" reported in pH pull–through studies[23,77,144]) and also the evaluation of pharmacological agents that stimulate or inhibit gastric secretion. The detection of secretion layering even under placebo conditions using a glucose meal with a known low secretion stimulating effect[128] underlines the sensitivity of the method and its potential usefulness for physiological measurement in health and disease.

Conclusion and Outlook

Conclusion

The projects presented in this thesis demonstrate the application of MRI for the assessment of human gastric motor function and gastric secretion and provide a deeper insight into the complex mechanisms controlling the processing and digestion of food in the gastrointestinal (GI) tract. Among various other measurement techniques such as scintigraphy, breath test, intraluminal manometry, ultrasound and esophageal pH–metry, MRI proves to be the optimal method for basic GI research of gastric motility disorders. MRI is noninvasive, free of ionizing radiation, provides excellent soft tissue contrast and has been shown to assess reliably gastric emptying, gastric accommodation and gastric peristaltic activity in humans.[29–31]

MRI for the Quantification of Gastric Secretion

The main project of this thesis aims on the development, evaluation and application of a MRI technique for the quantification of gastric secretion. The fundamental idea is based on calibrating the longitudinal relaxation time T_1 against the contrast agent concentration of a paramagnetic labeled liquid test meal. Since immediately after ingestion the labeled meal is diluted and mixed with gastric secretion, dynamic T_1 mapping together with the calibration allows the assessment of changes in gastric secretion volume.

Fast T_1 Mapping Technique

The fast T_1 mapping technique introduced in Chapter 1 is based on the variable flip angle approach (VFA).[58,59] This technique provides T_1 maps at high temporal resolution by acquiring multiple T_1 fast–field echo (T_1–FFE) images using different flip angles.[48] Optimization methods were developed to generate abdominal T_1 maps at high temporal (2.3 s per T_1 map) and spatial (2.8×2.8×15 mm^3) resolution with maximized accuracy over a wide T_1 range of 100 – 800 ms. The obtained resolution is sufficient to assess reliably the distribution and dynamics of orally or intravenously applied paramagnetic contrast agents.[56]

Maximization in T_1 accuracy was achieved by applying optimized flip angles, using RF excitation pulses with a high specific bandwidth to improve the slice excitation profile, ensuring effectual RF and gradient spoiling and steady state condition, and correcting flip angle variations due to inhomogeneous RF excitation. Deviations from the nominal flip angle represented the dominant source of error for T_1 determination and resulted in an underestimation of T_1 up to 10 % at 1.5 T and even up to 30 % at 3 T. Flip angle variations were corrected by performing an additional optimized fast flip angle imaging (B_1 mapping) technique based on a T_1–FFE sequence using two different, alternating repetition times.

Both T_1–FFE based T_1 and B_1 mapping techniques showed high sensitivity to motion. Breathing artifacts were minimized by performing measurements during breath–hold and flow artifacts of the liquid test meal by increasing its viscosity using locust bean gum powder. However, flow artifacts due to high–velocity pulsatile blood flow as occurring in the abdominal aorta and vena cava were still present on the images since measurements were not ECG triggered (in order to minimize the acquisition time). An overlap of these flow artifacts with the stomach were prevented by choosing the phase–encoding direction of the transverse image slices in anterior–posterior direction.

Fast T_1 Mapping to Quantify Gastric Secretion

In Chapter 2, quantification of gastric secretion volume after administration of a labeled viscous glucose solution (test meal) by applying the fast and optimized T_1 mapping technique was evaluated. Gastric secretion volume was quantified by calibrating the longitudinal relaxation time T_1 against the concentration of the Gadolinium (Gd) based paramagnetic contrast agent C_{Gd} of the diluted test meal and using basic Equation (2.1). Since dilution and mixing represent dynamic processes which alter continuously the macromolecular composition of the test meal, relaxivity r_1 and T_{10} remain not constant but change as a function of the macromolecular concentration C_M. In an *in vitro* experiment, this dependency was determined for the test meal that was continuously diluted with a 0.1 N HCl solution at physiological temperature of 37°C. Results showed a linear relationship between r_1 and C_M and between T_{10} and C_M. The 0.1 N HCl solution used for *in vitro* calibration represented an appropriate surrogate for gastric secretion because T_1 and pH values were observed to be similar to those of gastric juice (GJ). This was expected since HCl is the main component of GJ[87] and both solutions have a similar aqueous composition.

To validate the application of the derived *in vitro* calibration curve for *in vivo* quantification of gastric secretion volume, T_1 maps of four diluted test meals at specific dilution ratios were analyzed in five healthy volunteers. There was an excellent agreement between *in vivo* mean T_1 values and reference *in vitro* T_1 values. *In vivo* results further showed an increase in the width of the T_1 distribution for low C_{Gd} (long nominal T_1 values). This implies that quantification of gastric secretion volume using the fast and optimized T_1 mapping technique is effective in combination with test meals labeled with high contrast agent concentration ($\geqslant 200$ μM) and will be more inaccurate as contrast agent concentration decreases (< 200 μM).

Effect of Pentagastrin on Gastric Secretion

The novel MRI technique described in Chapters 1 & 2 was applied in twelve healthy volunteers to investigate the effect of gastric secretion on gastric physiology and emptying in the fasted and fed states, as presented in Chapter 4. Gastric secretion as stimulated by pentagastrin was compared to placebo in a randomized, double–blind study design. MRI volume scans were performed to assess total gastric volume (TGV) and gastric content volume (GCV) and combined T_1 and B_1 mapping scans to quantify volume and distribution of gastric secretion.

In the fasted state, GCV did not change during the placebo infusion. However, as expected, pentagastrin stimulated gastric secretion causing a continuous increase in GCV which resulted in a maximum at the end of the infusion phase of 95 – 171 ml. Thereafter, GCV decreased in a rapid, exponential manner similar to water.

In the fed state, gastric secretion and gastric emptying both commenced with meal ingestion, as evidenced by the development of a layer of highly diluted meal contents on top of the meal (observed on T_1 maps) and the presence of gastric contents in the duodenum, respectively. In the pentagastrin condition, TGV and GCV showed a pronounced early volume increase followed by a linear decrease which is typical for nutrient liquids. The rise in TGV and GCV was closely associated with the volume of the secretion layer forming on top of the meal. However, this increase was not observed during placebo infusion and the volume of the secretion layer was reduced. This finding indicates that the initial rate of gastric secretion was higher than the initial rate of gastric emptying for pentagastrin but not for placebo.

The dynamic change in TGV and GCV was similar in both pentagastrin and placebo condition. This suggests that gastric relaxation is driven by changes in GCV only and not by direct effects of gastrin on gastric wall tone. In addition, since maximal gastric emptying rate was not different between the two conditions, the significantly prolonged gastric half emptying time T_{50} during pentagastrin infusion was caused by the early increase in GCV. Thus, T_{50} as single summary endpoint often analyzed in physiological studies does not correctly reflect overall gastric emptying dynamics.

MRI for the Analysis of Gastric Motor Function

State–of–the–art clinical MRI systems are built in compact architecture restricting imaging to the lying body position. This is of major importance for gastric MRI since several studies using scintigraphy have shown that posture influences gastric function.[34-38,95]

In Chapter 3, an imaging protocol was developed and implemented on two MRI systems of different architecture and magnetic field strength (whole–body 1.5 T vs. open–configuration 0.5 T) allowing the simultaneous assessment of stomach, meal, and intragastric air volume; gastric relaxation and emptying; intragastric meal distribution; and peristaltic frequency and velocity in the right decubitus (RP) and seated body position (SP). In RP, post–prandial stomach geometry and meal distribution is most similar compared to the physiological upright body position because the distal stomach is always filled with gastric contents, and intragastric air is confined to the proximal stomach. The influence of posture on gastric secretion was not analyzed since T_1 values and thus $T_1 - C_{Gd}$ calibration and sensitivity to assess changes in C_{Gd} are different for 1.5 T and 0.5 T.

Results showed a slight difference in gastric emptying dynamics for the two body positions. Meal emptied faster in RP resulting in a small difference in remaining meal volume between RP (214 ml) and SP (257 ml) after 90 min. Altered gastric emptying dynamics was also reflected by the difference in intragastric air volume between RP and SP while stomach volume remained similar. Since gastric peristaltic frequency was similar for both positions and gastric peristaltic velocity did not correlate with the emptying dynamics, a change in intragastric pressure (induced by vagal activity) rather than a change in peristaltic activity was considered to cause the faster emptying in RP. Although similar results were observed in previous gastric emptying studies,[97,98,145] only the simultaneous measurement of gastric volume together with intragastric pressure within one imaging session will corroborate the hypothesis of pressure–driven gastric emptying.

However, observed meal volume difference after 90 min was much smaller compared to pathological changes in meal volume for patients suffering from gastric motility disorders.[96] This implies that gastric MRI performed in RP might become the method of choice for the assessment of GI function and motility disorders in humans.

Outlook

In the following, future research projects are presented aiming on further investigating basic physiology of human gastric secretion (in particular the effect of drugs to inhibit the production of gastric secretion and the structure and function of the gastroesophageal junction as key defence against acid reflux) and quantifying the dissolution and distribution of orally administered drugs in the small intestine.

Effect of proton pump inhibitors on gastric secretion

Besides of drugs stimulating the production of gastric secretion (e.g. pentagastrin), proton pump inhibitors (PPIs) are also of clinical interest. Basically, PPIs irreversibly block the hydrogen/potassium adenosine triphosphatase enzyme system (H^+/K^+ ATPase, commonly known as proton pump) of gastric parietal cells resulting in an inhibition of gastric secretion. Among a variety of other drugs, PPIs are the most effective inhibitors for treatment of acid–related disorders, as for example gastroesophageal reflux disease and peptic ulcer.[146]

Using the fast and optimized T_1 mapping technique introduced in Chapters 1 & 2, the effect of PPI (esomeprazole) on the volume and distribution of gastric secretory products in the GI tract may be investigated. In order to evaluate the efficacy of esomeprazole, a specific test meal acting as stimulus for gastric secretion must be applied. There are different constituents of food that provoke this stimulatory effect: peptic digests of protein, ethanol, coffee, and calcium.[147] However, in contrast to the results for pentagastrin (Chapter 4), differences in contrast agent concentration between esomeprazole and placebo (comparator drug) will be much smaller. As a consequence, the sensitivity of the fast T_1 mapping technique for accurate T_1 determination must be increased. This can be achieved by increasing the SNR using state–of–the–art receive coils and imaging at high magnetic field strengths (≥ 1.5 T). *In vitro* measurements on a 1.5 T whole–body MRI system performed by Philips Research Hamburg have shown that the SNR in an oval water phantom of 25 cm × 35 cm using a 32-channel cardiac coil is up to a factor of 1.5 higher compared to a standard 5-channel phased-array cardiac coil. The gain in SNR by imaging at high magnetic field strengths, however, is partly compensated by increased flip angle variations caused by B_1 field inhomogeneities.

Structure and Function of Gastroesophageal Junction

The gastroesophageal junction (GEJ) is of key importance in the defence against gastroesophageal reflux disease (GERD). Most reflux events occur during transient lower esophageal sphincter relaxations (TLESRs) in response to stomach distension. As shown in a variety of studies, factors other than GEJ function alone determine the risk of acid reflux during TLESR.[148–151] These factors include the structural integrity of the GEJ,[152–155] gastroesophageal pressure gradient (increased by straining and obesity),[156] meal volume relative to gastric capacity, and the distribution of gastric acid in the stomach.[23,157,158] However, standard measurement techniques allow only the assessment of some of the functional but none of the structural factors that protect the esophagus from reflux events.

Application of dedicated MR imaging techniques may allow to assess the interaction of structure and function at the GEJ and stomach that protects the esophagus from gastroesophageal reflux. Thereby, spatial and temporal image resolution must be maximized for accurate structural and functional assessment of the GEJ. Since reflux events do not occur periodically, a continuous image acquisition is necessary to reliably detect all reflux events. Continuous image acquisition and reconstruction, however, make high demands on the hardware of the MRI system (reconstructor). Therefore, high–resolution manometry pressure data may concurrently be recorded. The occurrence of a pressure wave induced by a reflux event on the manometry system may serve as trigger for manually starting dynamic data acquisition on the MRI system.

Quantification of Drug Distribution in the Small Intestine

Quantification of the dissolution and distribution of orally administered drugs in the small intestine is of crucial importance for the assessment of their efficacy and bioavailability. In recent studies, MRI has been used to qualitatively assess the release and distribution of a colloidal drug model containing Gd–DOTA from a gelatine capsule in the human stomach.[39,159] Image analysis was performed by manually selecting an intensity threshold to separate pixels with increased signal intensity (induced by Gd–DOTA) from pixels with lower signal intensity (containing no Gd–DOTA). However, since this analysis method is based on signal intensity, it is highly dependent on sequence parameters, B_1 homogeneity and sensitivity of the receive coils. The presented fast and optimized T_1 mapping technique offers the possibility to overcome these problems by mapping T_1 and thus quantifying the contrast agent concentration of the drug model.

In the small intestine, motor activity (frequency of peristaltic contraction waves) is much higher than in the stomach. Mean contraction frequency of the intestines is typically around eleven per minute[160] compared to three per minute of the stomach.[41] Increased activity results in an increase of flow (artifacts) of the ingested test meal. Therefore, temporal resolution of image acquisition using the fast and optimized T_1 mapping technique must be improved. This can be achieved by combining the fast T_1 mapping technique with accelerated imaging methods such as SENSE[32] or k–t BLAST and k–t SENSE.[33] Concurrently, a high spatial image resolution is also required to accurately localize the contrast agent in the wrinkles and projections of the small intestine. This technique may be validated by measuring simultaneously the plasma drug concentration over time.

An alternative approach for the quantification of intestinal drug dissolution and distribution is to label a drug model with ^{19}F and monitor its fate in the GI tract by ^{19}F–MRI. Fluorinated contrast agents have already been successfully applied in gastric ^{19}F–MRI for the assessment of gastric emptying and GI transit times.[161,162] Due to the lack of fluorine–containing metabolites in the human body, the ^{19}F signal can be assigned exclusively to the fluorinated contrast agent present in the GI tract. This means that the intensity of ^{19}F is a direct measure of the contrast agent concentration. The simultaneous acquisition of proton MR images allows to anatomically locate the detected ^{19}F signal. However, for ^{19}F–MRI specific hardware is required. Since the gyromagnetic ratio of the ^{19}F nuclei is different from that of protons (40.06 MHz/T vs. 42.58 MHz/T) custom–built send and receive coils tuned to the ^{19}F resonance frequency must be used.

References

[1] H. B. El-Serag and N. J. Talley. Health–related quality of life in functional dyspepsia. *Aliment Pharmacol Ther*, 18(4):387–93, 2003.

[2] T. M. Yunus, C. Mathis, K. M. Grabbe, L. J. Heinberg, and B. E. Lacy. Quality of life in patients with gastroparesis. *Am J Gastroenterol*, 98:S57, 2003.

[3] P. Pare, S. Ferrazzi, W. G. Thompson, E. J. Irvine, and L. Rance. An epidemiological survey of constipation in Canada: definitions, rates, demographics, and predictors of health care seeking. *Am J Gastroenterol*, 96(11):3130–7, 2001.

[4] I. M. Gralnek, R. D. Hays, A. Kilbourne, B. Naliboff, and E. A. Mayer. The impact of irritable bowel syndrome on health–related quality of life. *Gastroenterology*, 119(3):654–60, 2000.

[5] B. E. Lacy, V. Barghout, and D. Bauer. Gastroparesis. The impact on work and daily activities. *Am J Gastroenterol*, 100:S332, 2005.

[6] R. S. Sandler, J. E. Everhart, M. Donowitz, E. Adams, K. Cronin, C. Goodman, E. Gemmen, S. Shah, A. Avdic, and R. Rubin. The burden of selected digestive diseases in the United States. *Gastroenterology*, 122(5):1500–11, 2002.

[7] N. J. Talley, S. E. Gabriel, W. S. Harmsen, A. R. Zinsmeister, and R. W. Evans. Medical costs in community subjects with irritable bowel syndrome. *Gastroenterology*, 109(6):1736–41, 1995.

[8] M. I. Park and M. Camilleri. Gastroparesis: clinical update. *Am J Gastroenterol*, 101(5):1129–39, 2006.

[9] D. A. Drossman. The functional gastrointestinal disorders and the Rome III process. *Gastroenterology*, 130(5):1377–90, 2006.

[10] J. Tack, R. Bisschops, and G. Sarnelli. Pathophysiology and treatment of functional dyspepsia. *Gastroenterology*, 127(4):1239–55, 2004.

[11] C. Feinle-Bisset, R. Vozzo, M. Horowitz, and N. J. Talley. Diet, food intake, and disturbed physiology in the pathogenesis of symptoms in functional dyspepsia. *Am J Gastroenterol*, 99(1):170–81, 2004.

[12] J. Tack, N. J. Talley, M. Camilleri, G. Holtmann, P. Hu, J. R. Malagelada, and V. Stanghellini. Functional gastroduodenal disorders. *Gastroenterology*, 130(5):1466–79, 2006.

[13] A. O. Quartero, N. J. de Wit, A. C. Lodder, M. E. Numans, A. J. Smout, and A. W. Hoes. Disturbed solid–phase gastric emptying in functional dyspepsia: a meta–analysis. *Dig Dis Sci*, 43(9):2028–33, 1998.

[14] S. L. Lorena, E. Tinois, S. Q. Brunetto, E. E. Camargo, and M. A. Mesquita. Gastric emptying and intragastric distribution of a solid meal in functional dyspepsia: influence of gender and anxiety. *J Clin Gastroenterol*, 38(3):230–6, 2004.

[15] M. Simren, R. Vos, J. Janssens, and J. Tack. Unsuppressed postprandial phasic contractility in the proximal stomach in functional dyspepsia: relevance to symptoms. *Am J Gastroenterol*, 98(10):2169–75, 2003.

[16] N. J. Talley, M. D. Silverstein, L. Agreus, O. Nyren, A. Sonnenberg, and G. Holtmann. AGA technical review: evaluation of dyspepsia. *Gastroenterology*, 114(3):582–95, 1998.

[17] N. J. Talley. Dyspepsia. *Gastroenterology*, 125(4):1219–26, 2003.

[18] D. Armstrong. Gastroesophageal reflux disease. *Curr Opin Pharmacol*, 5(6):589–95, 2005.

[19] G. Holtmann, W. Siffert, S. Haag, N. Mueller, M. Langkafel, W. Senf, R. Zotz, and N. J. Talley. G–protein beta 3 subunit 825 CC genotype is associated with unexplained (functional) dyspepsia. *Gastroenterology*, 126(4):971–9, 2004.

[20] G. Holtmann, T. Liebregts, and W. Siffert. Molecular basis of functional gastrointestinal disorders. *Best Pract Res Clin Gastroenterol*, 18(4):633–40, 2004.

[21] F. K. Friedenberg and H. P. Parkman. Delayed gastric emptying: whom to test, how to test, and what to do. *Curr Treat Options Gastroenterol*, 9(4):295–304, 2006.

[22] D. K. Chitkara, M. Camilleri, A. R. Zinsmeister, D. Burton, M. El-Youssef, D. Freese, L. Walker, and D. Stephens. Gastric sensory and

motor dysfunction in adolescents with functional dyspepsia. *J Pediatr*, 146(4):448–50, 2005.

[23] J. Fletcher, A. Wirz, J. Young, R. Vallance, and K. E. McColl. Unbuffered highly acidic gastric juice exists at the gastroesophageal junction after a meal. *Gastroenterology*, 121(4):775–83, 2001.

[24] J. E. Pandolfino, Q. Zhang, S. K. Ghosh, J. Post, M. Kwiatek, and P. J. Kahrilas. Acidity surrounding the squamocolumnar junction in GERD patients: "acid pocket" versus "acid film". *Am J Gastroenterol*, 102(12):2633–41, 2007.

[25] J. Bratten and M. P. Jones. New directions in the assessment of gastric function: clinical applications of physiologic measurements. *Dig Dis*, 24(3-4):252–9, 2006.

[26] J. S. Lee, M. Camilleri, A. R. Zinsmeister, D. D. Burton, L. J. Kost, and P. D. Klein. A valid, accurate, office based non–radioactive test for gastric emptying of solids. *Gut*, 46(6):768–73, 2000.

[27] O. H. Gilja, J. G. Hatlebakk, S. Odegaard, A. Berstad, I. Viola, C. Giertsen, T. Hausken, and H. Gregersen. Advanced imaging and visualization in gastrointestinal disorders. *World J Gastroenterol*, 13(9):1408–1421, 2007.

[28] S. Emerenziani and D. Sifrim. New developments in detection of gastroesophageal reflux. *Curr Opin Gastroenterol*, 21(4):450–3, 2005.

[29] W. Schwizer, H. Maecke, and M. Fried. Measurement of gastric emptying by magnetic resonance imaging in humans. *Gastroenterology*, 103(2):369–76, 1992.

[30] P. Kunz, C. Feinle, W. Schwizer, M. Fried, and P. Boesiger. Assessment of gastric motor function during the emptying of solid and liquid meals in humans by MRI. *J Magn Reson Imaging*, 9(1):75–80, 1999.

[31] W. Schwizer, A. Steingoetter, M. Fox, T. Zur, M. Thumshirn, P. Boesiger, and M. Fried. Non–invasive measurement of gastric accommodation in humans. *Gut*, 51(Suppl 1):i59–62, 2002.

[32] K. P. Pruessmann, M. Weiger, M. B. Scheidegger, and P. Boesiger. SENSE: sensitivity encoding for fast MRI. *Magn Reson Med*, 42(5):952–62, 1999.

[33] J. Tsao, P. Boesiger, and K. P. Pruessmann. k–t BLAST and k–t SENSE: dynamic MRI with high frame rate exploiting spatiotemporal correlations. *Magn Reson Med*, 50(5):1031–42, 2003.

[34] M. Horowitz, K. Jones, M. A. Edelbroek, A. J. Smout, and N. W. Read. The effect of posture on gastric emptying and intragastric distribution of oil and aqueous meal components and appetite. *Gastroenterology*, 105(2):382–90, 1993.

[35] M. Anvari, M. Horowitz, R. Fraser, A. Maddox, J. Myers, J. Dent, and J. J. Jamieson. Effects of posture on gastric emptying of nonnutrient liquids and antropyloroduodenal motility. *Am J Physiol*, 268(5 Pt 1):G868–71, 1995.

[36] T. A. Spiegel, H. Fried, C. D. Hubert, S. R. Peikin, J. A. Siegel, and L. S. Zeiger. Effects of posture on gastric emptying and satiety ratings after a nutritive liquid and solid meal. *Am J Physiol Regul Integr Comp Physiol*, 279(2):R684–94, 2000.

[37] S. Doran, K. L. Jones, J. M. Andrews, and M. Horowitz. Effects of meal volume and posture on gastric emptying of solids and appetite. *Am J Physiol*, 275(5 Pt 2):R1712–8, 1998.

[38] J. G. Moore, F. L. Datz, P. E. Christian, E. Greenberg, and N. Alazraki. Effect of body posture on radionuclide measurements of gastric emptying. *Dig Dis Sci*, 33(12):1592–5, 1988.

[39] H. Faas, W. Schwizer, C. Feinle, H. Lengsfeld, C. de Smidt, P. Boesiger, M. Fried, and T. Rades. Monitoring the intragastric distribution of a colloidal drug carrier model by magnetic resonance imaging. *Pharm Res*, 18(4):460–6, 2001.

[40] A. Steingoetter, D. Weishaupt, P. Kunz, K. Mader, H. Lengsfeld, M. Thumshirn, P. Boesiger, M. Fried, and W. Schwizer. Magnetic resonance imaging for the in vivo evaluation of gastric–retentive tablets. *Pharm Res*, 20(12):2001–7, 2003.

[41] R. Treier, A. Steingoetter, D. Weishaupt, B. Marincek, P. Boesiger, M. Fried, and W. Schwizer. Gastric motor function and emptying in the right decubitus and seated body position as assessed by magnetic resonance imaging. *J Magn Reson Imaging*, 23(3):331–8, 2006.

[42] R. E. Hendrick and E. M. Haacke. Basic physics of MR contrast agents and maximization of image contrast. *J Magn Reson Imaging*, 3(1):137–48, 1993.

[43] M. T. Vlaardingerbroek and J. A. den Boer. *Magnetic resonance imaging – theory and practice*. Springer Verlag, Berlin, Heidelberg, New York, 2003.

[44] D. C. Look and D. R. Locker. Time saving in measurement of NMR and EPR relaxation times. *Rev Sci Instrum*, 41(2):250–1, 1970.

[45] P. Gowland and P. Mansfield. Accurate measurement of T1 in vivo in less than 3 seconds using echo–planar imaging. *Magn Reson Med*, 30(3):351–4, 1993.

[46] R. Deichmann, D. Hahn, and A. Haase. Fast T1 mapping on a whole–body scanner. *Magn Reson Med*, 42(1):206–9, 1999.

[47] P. Schmitt, M. A. Griswold, P. M. Jakob, M. Kotas, V. Gulani, M. Flentje, and A. Haase. Inversion recovery TrueFISP: quantification of T1, T2, and spin density. *Magn Reson Med*, 51(4):661–7, 2004.

[48] E. K. Fram, R. J. Herfkens, G. A. Johnson, G. H. Glover, J. P. Karis, A. Shimakawa, T. G. Perkins, and N. J. Pelc. Rapid calculation of T1 using variable flip angle gradient refocused imaging. *Magn Reson Imaging*, 5(3):201–8, 1987.

[49] P. L. Choyke and A. J. Dwyer. Functional tumor imaging with dynamic contrast–enhanced magnetic resonance imaging. *J Magn Reson Imaging*, 17(5):509–20, 2003.

[50] A. R. Padhani and J. E. Husband. Dynamic contrast–enhanced MRI studies in oncology with an emphasis on quantification, validation and human studies. *Clin Radiol*, 56(8):607–20, 2001.

[51] A. R. Padhani and M. O. Leach. Antivascular cancer treatments: functional assessments by dynamic contrast–enhanced magnetic resonance imaging. *Abdom Imaging*, 30(3):324–41, 2005.

[52] N. Grenier, F. Basseau, M. Ries, B. Tyndal, R. Jones, and C. Moonen. Functional MRI of the kidney. *Abdom Imaging*, 28(2):164–75, 2003.

[53] G. H. Hall, S. L. Atkin, and L. W. Turnbull. Use of dynamic contrast–enhanced MRI to assess the functional vascular pharmacokinetic parameters of normal human ovaries. *J Reprod Med*, 47(2):107–14, 2002.

[54] R. Materne, A. M. Smith, F. Peeters, J. P. Dehoux, A. Keyeux, Y. Horsmans, and B. E. Van Beers. Assessment of hepatic perfusion parameters with dynamic MRI. *Magn Reson Med*, 47(1):135–42, 2002.

[55] A. Steingoetter, P. Kunz, D. Weishaupt, K. Mader, H. Lengsfeld, M. Thumshirn, P. Boesiger, M. Fried, and W. Schwizer. Analysis of the meal–dependent intragastric performance of a gastric–retentive tablet assessed by magnetic resonance imaging. *Aliment Pharmacol Ther*, 18(7):713–20, 2003.

[56] A. Jackson. Analysis of dynamic contrast–enhanced MRI. *Br J Radiol*, 77 Spec No 2:S154–66, 2004.

[57] E. Kaldouidi and S. C. R. Williams. Relaxation time measurements in NMR imaging. *Concepts Magn Reson*, 5(3):217–42, 1993.

[58] R. K. Gupta. A new look at the method of variable nutation angle for measurement of spin–lattice relaxation times using Fourier transform NMR. *J Mag Res*, 25(1):231–5, 1977.

[59] J. Homer and M. S. Beevers. Driven–equilibrium single–pulse observation of T1 relaxation. A reevaluation of a rapid new method for determining spin–lattice relaxation times. *J Mag Res*, 63(2):287–297, 1985.

[60] H. L. Cheng and G. A. Wright. Rapid high–resolution T1 mapping by variable flip angles: accurate and precise measurements in the presence of radiofrequency field inhomogeneity. *Magn Reson Med*, 55(3):566–574, 2006.

[61] V. L. Yarnykh. Actual flip angle imaging in the pulsed steady state: a method for rapid three–dimensional mapping of the transmitted radiofrequency field. *Magn Reson Med*, 57(1):192–200, 2007.

[62] S. C. Deoni, B. K. Rutt, and T. M. Peters. Rapid combined T1 and T2 mapping using gradient recalled acquisition in the steady state. *Magn Reson Med*, 49(3):515–26, 2003.

[63] Y. Zur, M. L. Wood, and L. J. Neuringer. Spoiling of transverse magnetization in steady–state sequences. *Magn Reson Med*, 21(2):251–63, 1991.

[64] P. B. Roemer, W. A. Edelstein, C. E. Hayes, S. P. Souza, and O. M. Mueller. The NMR phased array. *Magn Reson Med*, 16(2):192–225, 1990.

[65] G. J. Stanisz, E. E. Odrobina, J. Pun, M. Escaravage, S. J. Graham, M. J. Bronskill, and R. M. Henkelman. T1, T2 relaxation and magnetization transfer in tissue at 3 T. *Magn Reson Med*, 54(3):507–12, 2005.

[66] E. Tadamura, H. Hatabu, W. Li, P. V. Prasad, and R. R. Edelman. Effect of oxygen inhalation on relaxation times in various tissues. *J Magn Reson Imaging*, 7(1):220–5, 1997.

[67] T. Q. Li and S. C. Deoni. Fast T1 mapping of the brain at 7 T with RF calibration using three point DESPOT1 method. In *Proceedings of the International Society for Magnetic Resonance in Medicine*, 2006.

[68] G. J. Parker, G. J. Barker, and P. S. Tofts. Accurate multislice gradient echo T1 measurement in the presence of non–ideal RF pulse shape and RF field nonuniformity. *Magn Reson Med*, 45(5):838–45, 2001.

[69] R. Venkatesan, W. Lin, and E. M. Haacke. Accurate determination of spin density and T1 in the presence of RF field inhomogeneities and flip angle miscalibration. *Magn Reson Med*, 40(4):592–602, 1998.

[70] R. Stollberger and P. Wach. Imaging of the active B1 field in vivo. *Magn Reson Med*, 35(2):246–51, 1996.

[71] R. Klinke. *Lehrbuch der Physiologie*. Thieme Verlag, Stuttgart, 2003.

[72] A. J. Gindea, J. Slater, and I. Kronzon. Doppler echocardiographic flow velocity measurements in the superior vena cava during the Valsalva maneuver in normal subjects. *Am J Cardiol*, 65(20):1387–91, 1990.

[73] D. I. Hoult. Solution of the Bloch Equations in the presence of a varying B1 field – approach to selective pulse analysis. *Journal of Magnetic Resonance*, 35(1):69–86, 1979.

[74] R. Treier, A. Steingoetter, O. Goetze, P. Boesiger, M. Fried, and W. Schwizer. Quantitative and non–invasive assessment of postprandial intragastric dilution as assessed by magnetic resonance imaging (MRI). *Gastroenterology*, 130(4):A194–A195, 2006.

[75] P. T. Holt and R. M. Russell. *Chronic Gastritis and Hypochlorhydria in the Elderly*. CRC Press, Stuttgart, 1993.

[76] S. Kwiecien and S. J. Konturek. Gastric analysis with fractional test meals (ethanol, caffeine, and peptone meal), augmented histamine or pentagastrin tests, and gastric pH recording. *J Physiol Pharmacol*, 54 Suppl 3:69–82, 2003.

[77] H. P. Simonian, L. Vo, S. Doma, R. S. Fisher, and H. P. Parkman. Regional postprandial differences in pH within the stomach and gastroesophageal junction. *Dig Dis Sci*, 50(12):2276–85, 2005.

[78] O. Goetze, A. Steingoetter, D. Menne, I. R. van der Voort, M. A. Kwiatek, P. Boesiger, D. Weishaupt, M. Thumshirn, M. Fried, and W. Schwizer.

The effect of macronutrients on gastric volume responses and gastric emptying in humans: a magnetic resonance imaging study. *Am J Physiol Gastrointest Liver Physiol*, 292(1):G11–7, 2007.

[79] L. Marciani, P. Manoj, B. P. Hills, R. J. Moore, P. Young, A. Fillery-Travis, R. C. Spiller, and P. A. Gowland. Echo–planar imaging relaxometry to measure the viscosity of a model meal. *J Magn Reson*, 135(1):82–6, 1998.

[80] L. Marciani, P. A. Gowland, R. C. Spiller, P. Manoj, R. J. Moore, P. Young, and A. J. Fillery-Travis. Effect of meal viscosity and nutrients on satiety, intragastric dilution, and emptying assessed by MRI. *Am J Physiol Gastrointest Liver Physiol*, 280(6):G1227–33, 2001.

[81] R. Treier, A. Steingoetter, M. Fried, W. Schwizer, and P. Boesiger. Optimized and combined T1 and B1 mapping technique for fast and accurate T1 quantification in contrast–enhanced abdominal MRI. *Magn Reson Med*, 57(3):568–76, 2007.

[82] G. J. Stanisz and R. M. Henkelman. Gd–DTPA relaxivity depends on macromolecular content. *Magn Reson Med*, 44(5):665–7, 2000.

[83] O. Goetze, M. Fried, and C. Gubler. Placement control by transabdominal ultrasound of duodenal feeding tubes: a feasible alternative for trials in a healthy volunteer study population. *Endoscopy*, 39:E309, 2007.

[84] D. W. Marquardt. An algorithm for least–squares estimation of nonlinear parameters. *SIAM J. Appl. Math.*, 11:431–41, 1963.

[85] M. A. Kwiatek, A. Steingoetter, A. Pal, D. Menne, J. G. Brasseur, G. S. Hebbard, P. Boesiger, M. Thumshirn, M. Fried, and W. Schwizer. Quantification of distal antral contractile motility in healthy human stomach with magnetic resonance imaging. *J Magn Reson Imaging*, 24(5):1101–9, 2006.

[86] L. Marciani, P. Young, J. Wright, R. J. Moore, D. F. Evans, R. C. Spiller, and P. A. Gowland. Echoplanar imaging in GI clinical practice: assessment of gastric emptying and antral motility in four patients. *J Magn Reson Imaging*, 12(2):343–6, 2000.

[87] M. L. Schubert. Gastric secretion. *Curr Opin Gastroenterol*, 19(6):519–25, 2003.

[88] N. Bloembergen, E. M. Purcell, and R. V. Pound. Relaxation effects in Nuclear Magnetic Resonance absorption. *Physical Review*, 73(7):679–712, 1948.

[89] I. Solomon. Relaxation processes in a system of 2 spins. *Physical Review*, 99(2):559–565, 1955.

[90] O. Goetze, R. Treier, A. Steingoetter, M. Fox, P. Boesiger, M. Fried, and W. Schwizer. Effect of gastric secretion on gastric volume responses, emptying and intragastric dilution in humans – a magnetic resonance imaging (MRI) study. *Gastroenterology*, 132(4):A580, 2007.

[91] I. N. Bronstein. *Handbook of mathematics*. Springer Verlag, Berlin, 2004.

[92] C. Feinle, P. Kunz, P. Boesiger, M. Fried, and W. Schwizer. Scintigraphic validation of a magnetic resonance imaging method to study gastric emptying of a solid meal in humans. *Gut*, 44(1):106–11, 1999.

[93] W. Schwizer, M. Fox, and A. Steingoetter. Non–invasive investigation of gastrointestinal functions with magnetic resonance imaging: towards an "ideal" investigation of gastrointestinal function. *Gut*, 52(Suppl 4):iv34–9, 2003.

[94] J. Borovicka, R. Lehmann, P. Kunz, R. Fraser, C. Kreiss, G. Crelier, P. Boesiger, G. A. Spinas, M. Fried, and W. Schwizer. Evaluation of gastric emptying and motility in diabetic gastroparesis with magnetic resonance imaging: effects of cisapride. *Am J Gastroenterol*, 94(10):2866–73, 1999.

[95] P. Boulby, P. Gowland, V. Adams, and R. C. Spiller. Use of echo planar imaging to demonstrate the effect of posture on the intragastric distribution and emptying of an oil/water meal. *Neurogastroenterol Motil*, 9(1):41–7, 1997.

[96] H. P. Parkman, W. L. Hasler, and R. S. Fisher. American Gastroenterological Association technical review on the diagnosis and treatment of gastroparesis. *Gastroenterology*, 127(5):1592–622, 2004.

[97] K. Indireshkumar, J. G. Brasseur, H. Faas, G. S. Hebbard, P. Kunz, J. Dent, C. Feinle, M. Li, P. Boesiger, M. Fried, and W. Schwizer. Relative contributions of "pressure pump" and "peristaltic pump" to gastric emptying. *Am J Physiol Gastrointest Liver Physiol*, 278(4):G604–16, 2000.

[98] T. Hausken, M. Mundt, and M. Samsom. Low antroduodenal pressure gradients are responsible for gastric emptying of a low–caloric liquid meal in humans. *Neurogastroenterol Motil*, 14(1):97–105, 2002.

[99] M. L. Malbrain. Different techniques to measure intra–abdominal pressure (IAP): time for a critical re–appraisal. *Intensive Care Med*, 30(3):357–71, 2004.

[100] G. S. Hebbard, K. Reid, W. M. Sun, M. Horowitz, and J. Dent. Postural changes in proximal gastric volume and pressure measured using a gastric barostat. *Neurogastroenterol Motil*, 7(3):169–74, 1995.

[101] T. Vybiral, R. J. Bryg, M. E. Maddens, and W. E. Boden. Effect of passive tilt on sympathetic and parasympathetic components of heart rate variability in normal subjects. *Am J Cardiol*, 63(15):1117–20, 1989.

[102] L. A. Lipsitz, J. Mietus, G. B. Moody, and A. L. Goldberger. Spectral characteristics of heart rate variability before and during postural tilt. Relations to aging and risk of syncope. *Circulation*, 81(6):1803–10, 1990.

[103] I. E. Hjelland, T. Hausken, S. Svebak, S. Olafsson, and A. Berstad. Vagal tone and meal–induced abdominal symptoms in healthy subjects. *Digestion*, 65(3):172–6, 2002.

[104] C. D. Kuo and G. Y. Chen. Comparison of three recumbent positions on vagal and sympathetic modulation using spectral heart rate variability in patients with coronary artery disease. *Am J Cardiol*, 81(4):392–6, 1998.

[105] G. Y. Chen and C. D. Kuo. The effect of the lateral decubitus position on vagal tone. *Anaesthesia*, 52(7):653–7, 1997.

[106] G. Tougas, M. Anvari, J. Dent, S. Somers, D. Richards, and G. W. Stevenson. Relation of pyloric motility to pyloric opening and closure in healthy subjects. *Gut*, 33(4):466–71, 1992.

[107] R. Heddle, D. Fone, J. Dent, and M. Horowitz. Stimulation of pyloric motility by intraduodenal dextrose in normal subjects. *Gut*, 29(10):1349–57, 1988.

[108] C. H. Malbert, C. Mathis, and J. P. Laplace. Vagal control of transpyloric flow and pyloric resistance. *Dig Dis Sci*, 39(12 Suppl):24S–27S, 1994.

[109] C. H. Malbert, C. Mathis, and J. P. Laplace. Vagal control of pyloric resistance. *Am J Physiol*, 269(4 Pt 1):G558–69, 1995.

[110] C. A. Paterson, M. Anvari, G. Tougas, and J. D. Huizinga. Determinants of occurrence and volume of transpyloric flow during gastric emptying of liquids in dogs: importance of vagal input. *Dig Dis Sci*, 45(8):1509–16, 2000.

[111] C. Mathis and C. H. Malbert. Erythromycin gastrokinetic activity is partially vagally mediated. *Am J Physiol*, 274(1 Pt 1):G80–6, 1998.

[112] R. Linke, W. Muenzing, K. Hahn, and K. Tatsch. Evaluation of gastric motility by Fourier analysis of condensed images. *Eur J Nucl Med*, 27(10):1531–7, 2000.

[113] S. D. Kuiken, M. Samsom, M. Camilleri, B. P. Mullan, D. D. Burton, L. J. Kost, T. J. Harsyman, B. H. Brinkmann, and M. K. O'Connor. Development of a test to measure gastric accommodation in humans. *Am J Physiol*, 277(6 Pt 1):G1217–21, 1999.

[114] R. J. Bennink, B. D. van den Elzen, S. D. Kuiken, and G. E. Boeckxstaens. Noninvasive measurement of gastric accommodation by means of pertechnetate SPECT: limiting radiation dose without losing image quality. *J Nucl Med*, 45(1):147–52, 2004.

[115] E. P. Bouras, S. Delgado-Aros, M. Camilleri, E. J. Castillo, D. D. Burton, G. M. Thomforde, and H. J. Chial. SPECT imaging of the stomach: comparison with barostat, and effects of sex, age, body mass index, and fundoplication. Single photon emission computed tomography. *Gut*, 51(6):781–6, 2002.

[116] D. Sifrim, R. Mittal, R. Fass, A. Smout, D. Castell, J. Tack, and H. Gregersen. Review article: acidity and volume of the refluxate in the genesis of gastro-oesophageal reflux disease symptoms. *Aliment Pharmacol Ther*, 25(9):1003–17, 2007.

[117] R. Dimaline and A. Varro. Attack and defence in the gastric epithelium - a delicate balance. *Exp Physiol*, 92(4):591–601, 2007.

[118] J. R. Malagelada, G. F. Longstreth, W. H. Summerskill, and V. L. Go. Measurement of gastric functions during digestion of ordinary solid meals in man. *Gastroenterology*, 70(2):203–10, 1976.

[119] J. R. Malagelada, V. L. Go, and W. H. Summerskill. Different gastric, pancreatic, and biliary responses to solid-liquid or homogenized meals. *Dig Dis Sci*, 24(2):101–10, 1979.

[120] M. Fried, E. A. Mayer, J. B. Jansen, C. B. Lamers, I. L. Taylor, S. R. Bloom, and J. H. Meyer. Temporal relationships of cholecystokinin release, pancreatobiliary secretion, and gastric emptying of a mixed meal. *Gastroenterology*, 95(5):1344–50, 1988.

[121] M. Feldman. Comparison of acid secretion rates measured by gastric aspiration and by in vivo intragastric titration in healthy human subjects. *Gastroenterology*, 76(5 Pt 1):954–7, 1979.

[122] R. Treier, A. Steingoetter, O. Goetze, M. Fox, M. Fried, W. Schwizer, and P. Boesiger. Fast and optimized t1 mapping technique for the noninvasive quantification of gastric secretion. *J Magn Reson Imaging*, in press, 2008.

[123] W. Schwizer, A. Steingoetter, and M. Fox. Magnetic resonance imaging for the assessment of gastrointestinal function. *Scand J Gastroenterol*, 41(11):1245–60, 2006.

[124] O. Goetze, D. Menne, M. A. Kwiatek, H. Fruehauf, A. Steingoetter, R. Treier, M. Fried, and W. Schwizer. Modeling of gastric volume data to assess gastric accommodation and emptying following ingestion of liquid meals. *Neurogastroenterol Motil*, 17:A97, 2005.

[125] J. D. Elashoff, T. J. Reedy, and J. H. Meyer. Analysis of gastric emptying data. *Gastroenterology*, 83(6):1306–12, 1982.

[126] J. C. Pinheiro and D. M. Bates. *Mixed–effects models in S and S-Plus*. Springer Verlag, Berlin, Heidelberg, New York, 2000.

[127] R development core team. *R reference manual - base package - volume 1*. Network Theory Limited, Bristol, 2003.

[128] C. T. Richardson, J. H. Walsh, M. I. Hicks, and J. S. Fordtran. Studies on the mechanisms of food-stimulated gastric acid secretion in normal human subjects. *J Clin Invest*, 58(3):623–31, 1976.

[129] B. G. Wilbur and K. A. Kelly. Gastrin pentapeptide decreases canine gastric transmural pressure. *Gastroenterology*, 67(6):1139–42, 1974.

[130] K. A. Kelly. Gastric emptying of liquids and solids: roles of proximal and distal stomach. *Am J Physiol*, 239(2):G71–6, 1980.

[131] J. E. Valenzuela and M. I. Grossman. Effect of pentagastrin and caerulein on intragastric pressure in the dog. *Gastroenterology*, 69(6):1383–4, 1975.

[132] A. L. Meulemans, A. L. Wellens, and J. A. Schuurkes. Gastric secretion but not nitric oxide is involved in pentagastrin-induced gastric relaxation in conscious dogs. *Neurogastroenterol Motil*, 9(1):49–54, 1997.

[133] B. Mearadji, J. W. Straathof, C. B. Lamers, and A. A. Masclee. Effect of gastrin on proximal gastric motor function in humans. *Neurogastroenterol Motil*, 11(6):449–55, 1999.

[134] I. M. de Zwart, J. J. Haans, P. Verbeek, P. H. Eilers, A. de Roos, and A. A. Masclee. Functional tumor imaging with dynamic contrast–enhanced magnetic resonance imaging. *Am J Physiol Gastrointest Liver Physiol*, 292(1):G208–14, 2007.

[135] J. N. Hunt and N. Ramsbottom. Effect of gastrin ii on gastric emptying and secretion during a test meal. *Br Med J*, 4(5576):386–7, 1967.

[136] J. N. Hunt and M. T. Knox. The slowing of gastric emptying by four strong acids and three weak acids. *J Physiol*, 222(1):187–208, 1972.

[137] M. P. Schwartz, M. Samsom, and A. J. Smout. Human duodenal motor activity in response to acid and different nutrients. *Dig Dis Sci*, 46(7):1472–81, 2001.

[138] H. P. Parkman, J. L. Urbain, L. C. Knight, K. L. Brown, D. M. Trate, M. A. Miller, A. H. Maurer, and R. S. Fisher. Effect of gastric acid suppressants on human gastric motility. *Gut*, 42(2):243–50, 1998.

[139] M. P. Schwartz, M. Samsom, and A. J. Smout. Chemospecific alterations in duodenal perception and motor response in functional dyspepsia. *Am J Gastroenterol*, 96(9):2596–602, 2001.

[140] T. D. Lewis, S. M. Collins, J. E. Fox, and E. E. Daniel. Initiation of duodenal acid-induced motor complexes. *Gastroenterology*, 77(6):1217–24, 1979.

[141] M. Verkijk, H. A. Gielkens, C. B. Lamers, and A. A. Masclee. Effect of gastrin on antroduodenal motility: role of intraluminal acidity. *Am J Physiol*, 275(5 Pt 1):G1209–16, 1998.

[142] J. J. Meier, M. A. Nauck, A. Pott, K. Heinze, O. Goetze, K. Bulut, W. E. Schmidt, B. Gallwitz, and J. J. Holst. Glucagon-like peptide 2 stimulates glucagon secretion, enhances lipid absorption, and inhibits gastric acid secretion in humans. *Gastroenterology*, 130(1):44–54, 2006.

[143] P. T. Regan, J. R. Malagelada, E. P. Dimagno, and V. L. Go. Reduced intraluminal bile acid concentrations and fat maldigestion in pancreatic insufficiency: correction by treatment. *Gastroenterology*, 77(2):285–9, 1979.

[144] A. Hila, H. Bouali, S. Xue, D. Knuff, and D. O. Castell. Postprandial stomach contents have multiple acid layers. *J Clin Gastroenterol*, 40(7):612–7, 2006.

[145] A. Steingoetter, M. Fox, R. Treier, D. Weishaupt, B. Marincek, P. Boesiger, M. Fried, and W. Schwizer. Effects of posture on the physiology of gastric emptying: a magnetic resonance imaging study. *Scand J Gastroenterol*, 41(10):1155–64, 2006.

[146] C. Scarpignato, I. Pelosini, and F. Di Mario. Acid suppression therapy: where do we go from here? *Dig Dis*, 24(1-2):11–46, 2006.

[147] T. Yamada, D. H. Alpers, L. Laine, N. Kaplowitz, C. Owyang, and D. W. Powell. *Textbook of Gastroenterology*. Lippincott Williams and Wilkins, Philadelphia, 2003.

[148] R. H. Holloway, M. Hongo, K. Berger, and R. W. McCallum. Gastric distention: a mechanism for postprandial gastroesophageal reflux. *Gastroenterology*, 89(4):779–84, 1985.

[149] N. J. Trudgill and S. A. Riley. Transient lower esophageal sphincter relaxations are no more frequent in patients with gastroesophageal reflux disease than in asymptomatic volunteers. *Am J Gastroenterol*, 96(9):2569–74, 2001.

[150] K. Iwakiri, Y. Hayashi, M. Kotoyori, Y. Tanaka, A. Kawakami, C. Sakamoto, and R. H. Holloway. Transient lower esophageal sphincter relaxations (TLESRs) are the major mechanism of gastroesophageal reflux but are not the cause of reflux disease. *Dig Dis Sci*, 50(6):1072–7, 2005.

[151] A. J. Bredenoord, B. L. Weusten, W. L. Curvers, R. Timmer, and A. J. Smout. Determinants of perception of heartburn and regurgitation. *Gut*, 55(3):313–8, 2005.

[152] L. D. Hill, R. A. Kozarek, S. J. Kraemer, R. W. Aye, C. D. Mercer, D. E. Low, and C. E. Pope. The gastroesophageal flap valve: in vitro and in vivo observations. *Gastrointest Endosc*, 44(5):541–7, 1996.

[153] P. J. Kahrilas. Anatomy and physiology of the gastroesophageal junction. *Gastroenterol Clin North Am*, 26(3):467–86, 1997.

[154] J. E. Pandolfino, G. Shi, J. Curry, R. J. Joehl, J. G. Brasseur, and P. J. Kahrilas. Esophagogastric junction distensibility: a factor contributing to sphincter incompetence. *Am J Physiol Gastrointest Liver Physiol*, 282(6):G1052–8, 2002.

[155] A. J. Bredenoord, B. L. Weusten, R. Timmer, and A. J. Smout. Intermittent spatial separation of diaphragm and lower esophageal sphincter

favors acidic and weakly acidic reflux. *Gastroenterology*, 130(2):334–40, 2006.

[156] J. E. Pandolfino, H. B. El-Serag, Q. Zhang, N. Shah, S. K. Ghosh, and P. J. Kahrilas. Obesity: a challenge to esophagogastric junction integrity. *Gastroenterology*, 130(3):639–49, 2006.

[157] S. Emerenziani, X. Zhang, K. Blondeau, J. Silny, J. Tack, J. Janssens, and D. Sifrim. Gastric fullness, physical activity, and proximal extent of gastroesophageal reflux. *Am J Gastroenterol*, 100(6):1251–6, 2005.

[158] D. Sifrim. Relevance of volume and proximal extent of reflux in gastro-oesophageal reflux disease. *Gut*, 54(2):175–8, 2005.

[159] H. Faas, A. Steingoetter, C. Feinle, T. Rades, H. Lengsfeld, P. Boesiger, M. Fried, and W. Schwizer. Effects of meal consistency and ingested fluid volume on the intragastric distribution of a drug model in humans – a magnetic resonance imaging study. *Aliment Pharmacol Ther*, 16(2):217–24, 2002.

[160] J. M. Froehlich, M. A. Patak, C. von Weymarn, C. F. Juli, C. L. Zollikofer, and K. U. Wentz. Small bowel motility assessment with magnetic resonance imaging. *J Magn Reson Imaging*, 21(4):370–5, 2005.

[161] R. Schwarz, M. Schuurmans, J. Seelig, and B. Künnecke. 19F-MRI of perfluorononane as a novel contrast modality for gastrointestinal imaging. *Magn Reson Med*, 41(1):80–6, 1999.

[162] R. Schwarz, A. Kaspar, J. Seelig, and B. Künnecke. Gastrointestinal transit times in mice and humans measured with 27Al and 19F nuclear magnetic resonance. *Magn Reson Med*, 48(2):255–61, 2002.

Südwestdeutscher Verlag für Hochschulschriften

Wissenschaftlicher Buchverlag bietet
kostenfreie
Publikation
von
Dissertationen und Habilitationen

Sie verfügen über eine wissenschaftliche Abschlußarbeit zu aktuellen oder zeitlosen Fragestellungen, die hohen inhaltlichen und formalen Anspruchen genügt, und haben **Interesse an einer honorarvergüteten Publikation?**

Dann senden Sie bitte erste Informationen über Ihre Arbeit per Email an: info@svh-verlag.de.

Unser Außenlektorat meldet sich umgehend bei Ihnen.

Südwestdeutscher Verlag für Hochschulschriften
Aktiengesellschaft & Co. KG

Dudweiler Landstr. 99
D – 66123 Saarbrücken

www.svh-verlag.de

Printed by Books on Demand GmbH, Norderstedt / Germany